A Cursing Brain?

A Cursing Brain?

. . .

The
Histories
of
Tourette Syndrome

HOWARD I. KUSHNER

Harvard University Press
Cambridge, Massachusetts
London, England
1999

Pages 291–294 constitute an extension of the copyright page.

Library of Congress Cataloging-in-Publication Data

Kushner, Howard I.
A cursing brain? : the histories of Tourette syndrome /
Howard I. Kushner.
p.   cm.
Includes bibliographical references and index.
ISBN 0-674-18022-4 (cloth : alk. paper)
1. Tourette syndrome—History.
I. Title.
RC375.K87   1999
616.8′3—dc21   98-38733

To Carol

Contents

Preface

In the early 1980s I came in frequent contact with a teenage neighbor who displayed an array of strange behaviors. Michael (a pseudonym) continually blinked his eyes and, at times, contorted his body and appeared to shrug his shoulders. His conversations were interrupted by sounds that resembled barks or grunts. Sometimes, Michael would repeat the final words of his sentences or phrases or, more annoying, the words of those with whom he was conversing. He seemed compelled to ask incessant questions and to make inappropriate remarks. One day Michael explained that his odd behaviors were caused by a neurological disease, called Gilles de la Tourette syndrome (TS). Although a historian of psychiatry and neurology, I had only a vague knowledge of this disorder and, given my psychoanalytic background, I was skeptical that Michael's symptoms were organic. I began to imagine a psychoanalytic scenario that tied Michael's symptoms to early childhood conflicts. I continued to observe Michael as his grunts were replaced by shouted curses. Michael's father reported that none of Michael's medications controlled the unwanted movements and cursing for a sustained period, and those that were most effective had intolerable side effects.

At the time I was working on a history of suicide, but resolved that in the future I would explore Michael's bizarre illness. The suicide study required bringing myself up to speed on recent research on the relationship between depressive disorders and the operation of neurotransmitter systems. Meanwhile, I was team-teaching graduate seminars on the history of neurobiology. I began to see connections rather than contradictions between psychogenic theory and neural mechanisms. My evolving appreciation and understanding of these issues were reflected in the subtitle of my 1989 book, *A Psychocultural Biology of American Suicide.*[1] By the early 1990s, when I began researching the history of Tourette syndrome, I came equipped with a new set of skills informed by a decade's immersion in neurochemistry and neurobiology. I regularly was attending the brain cutting seminar for neuropathology residents at UCSD medical school. Like many others, my skepticism about psychoanalytic claims grew as I learned more about how the brain operates.

From what I had learned it seemed that the involuntary movements and vocalizations associated with Tourette syndrome were most likely tied to the transmission of the neurotransmitter dopamine in part of the brain known as the basal ganglia, a set of brain structures that regulate motor movements. One explanation was that neuronal connections in the basal ganglia were hypersensitive to dopamine. Increasing evidence pointed to an inherited (genetic) factor for this reaction, but just how these mechanisms might work, or if more than one gene was responsible, remained to be demonstrated.

Although my mind was now "prepared" to understand the possible organic factors of Tourette's, I was puzzled by a recurrent theme that emerged in my historical research. Throughout the nineteenth century, physicians had reported a persistent connection between symptoms similar to those displayed by Michael and a previous rheumatic fever illness. Rheumatic fever results when antibodies responding to the invading bacteria (Group A beta hemolytic streptococcal or GABHS) mistake a person's own cell proteins on the heart lining for those of the invading bacteria. This process is called "molecular mimicry" because a person's own molecules (called epitopes) mimic those of invader bacteria, resulting in antibodies binding not only with proteins on the invading bacteria, but also on one's own cells. The movement disorders that often seemed to follow these rheumatic attacks were generally classified as "choreas," the Greek word for dance. One of these choreas—Sydenham's—is now understood to result from a mechanism similar to that involved in rheumatic fever. In Sydenham's the activated antibodies attach themselves to part of the basal ganglia, interfering with the normal brain signaling. Thus, the involuntary movements of Sydenham's result from the downstream effects of a bacterial (GABHS) infection.

I wondered why no one had investigated a similar molecular

mechanism for Tourette's. Although neurologists are taught to distinguish Sydenham's from Tourette's on the basis of differences in symptom presentation and time course of the illness, clinical experience is less well demarcated than textbook and classroom descriptions. Often what appears first as Sydenham's (for instance, tongue protrusions) mutates into the movements associated with TS. In any case, slightly different sets of symptoms can emerge from the same underlying causes.[2]

I decided to apply for a grant to explore these possible connections from a historical perspective and sent a copy of my proposal to Sue Levi-Pearl, the scientific liaison of the Tourette Syndrome Association. Pearl put me in contact with Dr. Louise Kiessling, chief of pediatrics at Brown University Medical School's Memorial Hospital of Rhode Island, who was investigating the relationship between GABHS infections and TS. Dr. Kiessling invited me to participate in their research project. Thanks to a National Endowment for the Humanities, Science, and Technology grant, I spent two months in 1994 and 1996 at Kiessling's child developmental clinic. Under supervision and with patient permission, I sat in on examinations, family interviews, and generally participated in team discussions of cases. Along with her colleague Dr. Joseph Hallett, Kiessling explained how to look for protein bindings, discussed the pros and cons of assays (chemical analyses), and led me into the maze of rat research. My historical research supplemented the Brown team's investigations suggesting the possibility that at least some of those diagnosed with TS experienced onset and exacerbations of their symptoms subsequent to GABHS infections. Thus some, but not all, Tourette's cases might be enabled by a mechanism similar to that elaborated for Sydenham's.[3]

Since 1995 Dr. Kiessling and I have collaborated on clinical articles and presented papers at professional meetings outlining the utility of historical research for clinical investigations and vice versa.[4] I continue to be involved with the Brown research group and hope that their work, when combined with parallel investigations at the National Institutes of Health and other sites, will uncover the mechanisms resulting in TS for some of the afflicted. We also hope that these findings will lead eventually to robust interventions aimed at the causes rather than the symptoms of these behaviors. Because none of the research teams is persuaded that everyone with Tourette syndrome has been subject to a cross-reaction to a previous infection, the Brown team and others have speculated that there are several possible organic routes resulting in the group of symptoms now categorized as Tourette syndrome. If this turns out to be so, what we now refer to as Tourette syndrome most likely will be recategorized as a series of childhood movement disorders that look very much alike in symptom presentation, but that have identifiably distinct sets of causes. Thus, a team at the National Institutes of Mental Health has labeled this postinfectious scenario as PANDAS, or "pediatric autoimmune neuropsychiatric disorders associated with streptococcal infections."[5] If these distinctions prove resilient, postinfectious cases of Tourette's will be treated differently from those cases in which GABHS antibodies are absent.[6]

Although my involvement in this clinical research has influenced my thinking about the nature of Tourette syndrome, I am also a historian of medicine. As a historian, I am constantly reminded that one day's medical certainty gives way to more compelling explanations and interventions. Nowhere has this been more true than in the history of Tourette syndrome.

A Cursing Brain? traces the categorization and treatment of motor and vocal tics and their allied symptoms from the early nineteenth century to the present. An examination of the reports of physicians and those afflicted uncovers not only changing explanations for the causes of these behaviors, but also the extent to which the etiological (causative) theories of practitioners were sometimes authorized and internalized by the afflicted. Even though in retrospect this history uncovers numerous therapeutic failures, it also reveals that physicians and many of their patients reported successful interventions based on medical constructions that current medical science would find completely wrong-headed. Although from today's perspective the interventions followed may appear fraudulent, those who employed them were, for the most part, sincere and well-trained by contemporary standards. Moreover, there is ample evidence from many patients' case histories to support the effectiveness of a number of seemingly persuasive, yet contradictory, interventions. Thus, while I cannot deny my attraction to the molecular mimicry scenario, I continue to approach it with the skepticism that informs my attitude toward all tentative claims. It is in this spirit that I have undertaken the research and writing of this book.

Note on Terms

Various authors have used slightly different terms when referring to Gilles de la Tourette syndrome. Unless quoting directly from a source, "Tourette syndrome" and "Gilles de la Tourette syndrome" are used interchangeably to indicate the disorder described since the late 1960s, first by Arthur K. Shapiro and his colleagues and, in the 1980s, in the American Psychiatric Association's *Diagnostic and Statistical Manual of Mental Disorders* (the DSM-III and IV). "Tourette's" (with an "'s") is used synonymously for the first two terms when the word "syndrome" is omitted. The term "Gilles de la Tourette's disease" (or "illness") refers specifically to the classifications erected by Georges Gilles de la Tourette in 1885, which his mentor Jean-Martin Charcot labeled "maladie des tics de Gilles de la Tourette." Because "maladie" most closely translates into English as disease or illness, it should not be translated as "syndrome," which in French is also "syndrome." As readers of this book will learn, whether one views this disorder as a disease or as a syndrome is no trivial matter.

Except when quoting directly from a source, the modern spelling "ticcer" is used as synonymous with, and in place of, the older British and French spelling "tiqueur" or "tiquer." All forms, including also "ticker," are used by authors to mean involuntary motor movements.

A Cursing Brain?

An Elusive Syndrome

$$I$$

··· · ◆ ◆ ◆ · ···

Seated toward the back of a meeting room in a local hospital I listen to a speaker, a published writer in his early thirties. He periodically turns his head almost rhythmically, first to the left and then to the right, each time forming his lips as if to spit without actually expectorating. Behind me, a young man shouts out "Fuck you! I love you." A half minute later he bellows, "Sit down, shut up!" Neither the speaker nor any of the fifty or so persons in the audience reacts to any of this behavior. During the question period following the talk a woman sitting in the front row asks the speaker a question, which he answers; but, in the middle of his response he blurts out "Douche bag!" and then continues speaking to the woman as if the sounds had come from somewhere and someone else. She appears neither offended nor surprised by this outburst. Another hand is raised in the audience and for half an hour the speaker deftly handles the questions, always punctuating his responses with parenthetical foul language or remarks that seem so particularly inappropriate as to be purposeful. Thus, responding to a question from a noticeably obese woman, he interjects "Fat pig!" into his remarks. Meanwhile, the man in the back of the room continues to shout obscenities, as if competing for attention. Still, no one in the audience seems offended or even appears to notice the bizarre and offensive behavior of the speaker or his interlocutor.

The speaker who had been introduced at the outset as having a disorder called Gilles de la Tourette syndrome (TS) was addressing a support group comprised mainly of parents of diagnosed children.[1] The man in the audience shouting obscenities has been afflicted since his early teens, and as his neighbor for fifteen years, I had watched his symptoms develop from uncontrolled eye blinking to more pronounced facial and body tics accompanied by vocalizations that at first sounded like a person muttering to himself. When "Michael," as I will call him, approached his late teens, he began to curse and blurt out uncontrollable inappropriate remarks that made it difficult for him to socialize with peers or with anyone unaware of the nature of his offensive behavior. Even ordinary tasks became dangerous. Once, phoning an airline to make a reservation, Michael blurted out, "There's a bomb on the plane!" The next day the FBI appeared at his door to question a "suspected terrorist." Sometimes, when Michael is introduced to an African-American, he cannot stop himself from exclaiming, "Nigger!"

The speaker recounts similar experiences, one of which resulted in his removal from an airplane after he failed to abide a flight attendant's request that he cease cursing. His explanation that he suffered from a neurological disorder known as Tourette syndrome did not win sympathy from the attendant or other passengers. Michael, likewise, has been evicted from two apartments, has been served restraining orders barring him from local fast-food restaurants, and has been involved in litigation as a result of a landlord's complaint that Michael allegedly threatened to "drop a bowling ball from an airplane" on the man. Like the speaker, Michael attempts to explain to those with whom he comes into contact that his cursing and strange vocalizations are the result of Tourette syndrome. Nevertheless, the content and timing of these outbursts can be discomforting to anyone who is the target or for those who witness them. On the one hand, they are apt but painful characterizations, such as blurting out what may be obvious, but tasteless or tactless; for instance, exclaiming to an obese person that they are "fat." On the other hand, these unrestrained words and phrases can be threatening, as when Michael interjects "I want to rape you" when talking to a woman.

Such reactions influenced a local municipal court judge's decision to uphold an order restraining Michael from visiting a neighboring apartment complex. The judge refused to believe that a disease could account for Michael's cursing and threatening outbursts. At a subsequent hearing only the presence of a television news crew persuaded the judge to moderate his order. The judge's good will was, no doubt, stretched thin by Michael's periodic outburst of "I'll kill you, I love you!"

It is true that Michael has difficulty controlling his temper; thus, when particularly frustrated by the effects of his behavior or others' reactions to it, Michael sometimes will punch or kick a wall or damage fixtures in his apartment. Michael has never physically harmed others. However, many parents report that their afflicted children do sometimes hit, punch, and kick their siblings and their pets. While not apparent in either Michael or the speaker, other florid symptoms beyond cursing, which is clinically referred to as "coprolalia," include "copropraxia," the acting out of explicitly sexual gestures or displays; "echolalia," repetition of one's own or others' words or phrases; and "echopraxia," imitation of others' behaviors or actions.

These are the florid symptoms. Most persons diagnosed with Gilles de la Tourette syndrome neither curse nor display inappropriate sexual behavior, although they often have difficulty controlling their tempers. Typically, sufferers develop involuntary motor movements, referred to as "tics," during early childhood (ages seven to nine). The motor tics, which occur frequently throughout the day, generally involve head and neck jerking, eye blinking, tongue protrusions, shoulder shrugs, and various torso and limb movements. A diagnosis of Tourette syndrome requires one or more vocal tics to have been present for some time. These may include barks, grunts, yelps, and coughs. Tics and vocalizations appear suddenly and characteristically are rapid, recurrent, non-rhythmic, and stereotyped. Often these symptoms are coupled with obsessive and compulsive behaviors, such as a repeated series of actions that must be performed before entering or leaving a room. Tics wax and wane, often increasing in frequency and complexity, with later tics replacing earlier ones.[2] Motor movements, however, can be quite debilitating and painful. For instance, aside from the muscle strain caused by severe head-jerking, tics can make reading an arduous, if not impossible task.[3] Sometimes the tics disappear completely and never recur. Often, however, they merely remit, returning later in slightly different form with renewed force.

Commonly, parents of children with TS report their long and difficult journey through a medical maze in search of a diagnosis for their children's behavior. It is not at all unusual that the diagnosis is finally obtained when parents, armed with literature from newspapers, magazines, or from the Tourette Syndrome Association, educate their physicians about the syndrome. Resistance to labeling TS behavior as a disease, however, flows as much from the nature of its most florid symptoms as from a general lack of awareness. Like the municipal judge in Michael's case, even physicians, not least of all psychiatrists, have been reluctant to diagnose seemingly obvious antisocial behavior, such as inappropriate cursing or explicitly sexual displays, as an organic disorder. The controversy over whether these behaviors constitute an illness has a long history. Exploration of that history reveals much about the continuing reluctance to validate inappropriate social behaviors as having organic causes.

Once thought to be a rare occurrence, Tourette syndrome represents one of the fastest growing diagnoses in North America, with a prevalence of from 2.9 to 5.2 per 10,000.[4] A British team recently reported that 3 percent of thirteen- to fourteen-year-old secondary schoolchildren (299 per 10,000) fit the *Diagnostic and Statistical Manual of Mental Disorders-III-R* (DSM-III-R) classification of Tourette syndrome, which is four times higher than the previous highest rates reported in 1990.[5] All studies find that males are four to five times more likely to become afflicted than females. One recent investigation reported that among boys the prevalence may be as high as 5 per 100.[6] Even at the low end, estimates would make the prevalence of Tourette syndrome more than double that of teen suicide in the United States.[7]

Part of the reason for contradictory findings about prevalence is attributable to physicians' and researchers' disagreements about which symptoms to include and which to exclude in their definitions of Tourette's. That is because TS is a *syndrome* rather than a *disease*. Although the term "disease" can mean many things, including simply "distress," a disorder generally only graduates from its status as a syndrome when its underlying pathological causes or "etiologies," as complex as they may be, are uncovered. Measles, polio, smallpox, and sickle-cell anemia are labeled as diseases and not as syndromes because a tentative diagnosis based on signs and symptoms is confirmed or rejected through a laboratory test that indicates infection by a pathogen or the presence of an enabling gene. In contrast, the cause of a syndrome remains unknown.[8] With syndromes, like Tourette's, schizophrenia, and chronic fatigue, diagnoses depend on identification of a list of possible combinations of signs and symptoms that a person must display over an assigned period of time.[9] But as Terra Ziporyn notes, this list may vary from physician to physician: "One researcher, for example, may have defined chronic fatigue syndrome as including anyone who feels tired most of the time, while her colleague may have restricted it to persistent fatigue coupled with flu symptoms and perhaps even associated with a virus . . . Without an agreed-on definition, more sophisticated correlations and predictions become meaningless. If you don't know what chronic fatigue syndrome is, for example, how can you say what happens to people with chronic fatigue syndrome?"[10] The signs and symptoms that have been grouped together as Tourette's often seem to fit a spectrum, in which a single cause can result in a series of different symptoms. Alternatively, because its underlying pathology remains unknown, these signs and symptoms could result from a variety of disparate causes.[11]

Grouping certain symptoms, but not others, under a single "name" or category inevitably influences practitioners' and patients' understanding of the possible causes and course of a disorder. The purpose of syndrome construction is to focus a practitioner's view of a patient's illness by privileging certain symptoms while downplaying others. What physicians believe constitute the legitimate symptoms of a disorder can have a profound influence on a sufferer's experience. Certainly it influences which symptoms become the focus for treatment and which are viewed as unrelated to the syndrome.

Since the late nineteenth century, when Tourette's was initially described, psychiatrists have disagreed, often vigorously, over which symptoms and behaviors should be included as part of TS. These controversies continue. Beginning in the 1980s a serious dispute arose among experts over whether obsessive-compulsive behaviors should be included within the Tourette's typology.[12] More recently a division has emerged among those who believe that conduct disorders are part of TS and those who resist too wide a spectrum of symptoms.[13] In addition, clinicians disagree over the length of time a symptom must be present to be considered for a diagnosis and whether or not onset must occur before the age of eighteen.

Although one of the purposes of the American Psychiatric Association's diagnostic handbook, the DSM, is to provide uniform definitions of psychiatric syndromes, a comparison of the definitions and classifications of Tourette syndrome in the revisions of DSM (III-R) reveals that there has been no absolute agreement over time about what symptoms are necessary or how long they must persist for a diagnosis of Tourette syndrome.[14] A new effort at a uniform definition was made in the construction of Tourette's categories in the DSM-IV (1994),[15] but dissension over the typology erupted among both members of the committee that had been commissioned to write this section, as well as among other experts as soon as the volume appeared.[16] By 1997 a team of experienced Tourette's researchers at Yale University Medical School's Child Study Center admitted that continuing disagreements over what constitutes the symptoms that are part of TS (its phenotype) continue to frustrate attempts to locate its underlying pathogenesis (causes).[17]

Nevertheless, the vast majority of researchers are persuaded that, no matter what its constituents, Tourette's is an organic disorder. The most persuasive evidence for a physiological basis of Tourette's is that motor tics and involuntary vocalizations can often be controlled by drugs that act to suppress the transmission of the neurotransmitter dopamine in that part of the brain called the basal ganglia, which is responsible for certain motor movements. However, these drugs can only provide relief from the symptoms of Tourette's and do not eliminate the causes of the tics and vocalizations. Researchers continue to search for the underlying causes, which may reside in a genetic malfunction or an autoimmune reaction to a previous infection, or a combination of these and other cofactors.

Even as persuasive evidence mounts tying specific ticcing and obsessive-compulsive symptoms to organic mechanisms, a number of clinicians have resisted a biologically reductionist explanation of Tourette's. They point to the elusive quality of many symptoms that wax and wane for no apparent reason, sometimes disappearing completely. Because some of those afflicted are able to suppress their tics and vocalizations, releasing them minutes later away from the observation of others, skeptics are unpersuaded that TS behaviors are completely involuntary. Alternatively, these temporary suppressions could be understood as elaborate strategies, as Michael has attempted, substituting less offensive words when unable to resist the urge to curse. Oliver Sacks's portrayal of a ticcing drummer, "Witty Ticcy Ray," suggests that some are able to turn their symptoms into socially sanctioned behaviors.[18] But Sacks's description of a Canadian surgeon with florid TS symptoms, who is able to suspend his tics while performing operations or when flying his airplane, no doubt raises questions in many minds about the assertions that these behaviors are unintentional.[19]

Citing these and other examples of the seeming ability of some diagnosed with Tourette's to control their behaviors, a few observers insist that psychological conflicts play a crucial role in the onset and perpetuation of ticcing symptoms. Claiming that psychological factors influence the way that biochemistry and brain circuitry operate, this minority continues to frame physiological symptoms as manifestations of psychological conflicts.[20] Most often those who resist organic explanations point to cursing as evidence in support of their conclusions.

Although cursing is not present in every case of TS, this symptom, more than any other, periodically has brought Gilles de la Tourette syndrome to medical and popular notice. In fact, this odd behavior had intrigued researchers years before the disorder had a name and brought TS to the attention of those late nineteenth-century psychiatrists and neurologists who defined it as a separate disorder. It seems bizarre on the face of it, that something as rooted in culture as the utterance of inappropriate phrases or obscene words could be attached to organic disease. What in some societies are viewed as outrageous curses are seen as inoffensive in others. Moreover, even within the same societies words lose their offensive connotations over time.[21] What is most interesting about coprolalia in Tourette's sufferers is that they invoke the most inappropriate curses of their particular times and cultures.

Since physicians first noticed involuntary swearing in the mid-nineteenth century, there has been one widely acknowledged organic pathology connected with it. Broca's aphasia, a stroke in the part of the brain connected with articulation of speech, often is accompanied by reflexive cursing. Today, neurologists recognize a connection between destruction of normal speech and, as one contemporary researcher put it, the resultant "ability to swear like a sailor."[22] While the discovery of cursing behavior of aphasics was made simultaneously with that of convulsive ticcers, no link was obvious because no gross anatomical brain cell destruction was evident among ticcing coprolalics. Recently, however, a number of researchers have argued that, like aphasics, swearing Touretters also suffer from an alteration of those areas of the brain that inhibit uncontrolled emotional behavior.[23]

Even if neurobiology and biochemistry play an important role, a brain cannot "curse" without knowledge about what a culture views as a linguistic taboo. One of the striking features of those who display florid vocalization symptoms is a compulsion to articulate the most outrageous words for the occasion. It is as if, explains Tourette sufferer and author Adam Ward Seligman, someone comes to your door demanding, "don't think about monkeys." That command, of course, results in your thinking about nothing else. "Having Tourette syndrome," writes Seligman, "is a lot like not thinking about monkeys. The monkeys are the tics, vocalizations, urges, obsessions, behaviors and enactments that are with us constantly, overwhelming our daily lives. To live and function we have to keep the Tourette syndrome at bay—we have to try not to think about monkeys."[24]

Like Seligman's essays, explanations of what we today label "Tourette syndrome" have been and continue to rely on patients' stories. Unlike Seligman's, most of these stories have been collected and interpreted by clinicians. Diagnosing clinicians are influenced not only by a patient's story (often supplemented by reports from teachers, parents, and other family members), but also by the meanings given to it by the reports of referring physicians. When a patient returns for follow-up treatments, clinicians often discover how difficult it is to interpret their own, let alone their colleagues', earlier recording of the patient/family interview. On subsequent visits, they find themselves rewriting the earlier observations, giving new meaning to what was seen before. This scenario is even more complicated in the diagnosis and treatment of Tourette syndrome because of the changing nature of the symptoms and their tendency to wax and wane.

If clinicians face a daunting task interpreting the meaning of the symptoms of a patient they actually have examined, it is even more difficult to judge a published case, whose meaning has been interpreted in order to make a clinical argument about underlying causes or appropriate treatments. In its published form the clinical case becomes evidence for the author's conclusions. Even the most rigorous and honest reporters edit their presentations, often leaving out what previously might have been seen as a central element of the disorder (perhaps familial conflict) or what later might become valuable (for instance, evidence of recurrent bacterial infection).

As a result, the interpretation of clinical cases both influences and reflects clinicians' identification and treatment of Tourette syndrome. It also frames the way individuals have experienced this affliction. For, no matter what its underlying biology, how those with Tourette syndrome have experienced involuntary movements and unwelcome vocalizations has depended to a great extent upon culturally sanctioned expectations. A history of Tourette syndrome must, therefore, simultaneously explore three distinct but overlapping stories: that of the claims of medical knowledge, that of patients' experiences, and that of cultural expectations and assumptions.

These issues are most clearly illustrated by the multiple and competing readings of the most cited case of Tourette syndrome, that of the Marquise de Dampierre (1799–1884).[25] Even before the syndrome had a name, this cursing and ticcing nineteenth-century noblewoman served as an emblematic example for a variety of competing theories about the mechanisms associated with humans' ability to self-consciously control their actions and behaviors. Her behavior would be cited by Gilles de la Tourette as the most complete example of the disorder he described. Today, the story of the Marquise de Dampierre has become the conventional tale that appears in the opening paragraphs of almost every clinical overview of Tourette syndrome. A reinvestigation of that story is an essential starting point for our historical inquiry.

The Case of the Cursing Marquise

2

In August 1884 Parisian newspapers reported the death of the eighty-five-year-old Marquise de Dampierre, notorious for publicly shouting out, in the middle of conversations, inappropriate or obscene words, especially "shit and fucking pig."[1] Dampierre's bizarre behavior had been the subject of gossip for over half a century. Her story was originally reported in 1825 by Jean Marc Gaspard Itard (1775–1838), chief physician at l'Institution Royale des Sourds-muets in Paris, who, twenty-five years earlier, had gained worldwide notice as the physician charged with educating Victor, the so-called Wild Child of Aveyron.[2] Itard recorded the behavior of the then twenty-six-year-old Madame de Dampierre in an article published in *Archives Générales de Médecine*. "In the midst of a conversation that interests her extremely," Itard reported, "all of a sudden, without being able to prevent it, she interrupts what she is saying or what she is listening to with bizarre shouts and with words that are even more extraordinary and which make a deplorable contrast with her intellect and her distinguished manners. These words are for the most part gross swear words and obscene epithets and, something that is no less embarrassing for her than for the listeners, an extremely crude expression of a judgment or of an unfavorable opinion of someone in the group." The more she was revolted by a word's "grossness," explained Itard, "the more she is tormented by the fear that she will utter them, and this

Figure 2.1 In 1825 Jean Itard recorded the ticcing and cursing behavior of the twenty-six-year-old Marquise de Dampierre. Georges Gilles de la Tourette selected Itard's case of the marquise as his primary example in 1885 of the disease he labeled "maladie des tics."

preoccupation is precisely what puts them at the tip of her tongue where she can no longer control it."[3]

Sixty years after Itard published this description, a twenty-eight-year-old Parisian neurologist, Georges Gilles de la Tourette, selected this case as his first example of the illness that he classified as "maladie des tics."[4] The emblematic status of the case was certified by the assertion that in her later years the marquise had been diagnosed and treated by Gilles de la Tourette's mentor, Jean-Martin Charcot, chief physician of the Salpêtrière Hospital and the foremost neurologist of late nineteenth-century France. It was, in fact, at Charcot's explicit direction that Gilles de la Tourette had organized his and Charcot's clinical cases for publication.[5] And it was Charcot who almost immediately renamed the convulsive tic illness in honor of Gilles de la Tourette.

Virtually every general medical and popular overview of Tourette syndrome since has begun, as if by convention, with the case of the Marquise de Dampierre.[6] By the mid-1990s, the conventional citation of the case of the Marquise de Dampierre had evolved into an emblematic narrative, a sort of short-hand description of the onset and course of the disorder itself.[7] "Tourette's syndrome," wrote Stanley Finger in his encyclopedic *Origins of Neuroscience,* "can best be appreciated by looking directly at his [Gilles de la Tourette's] case of Madame de Dampierre."[8]

The marquise's story provides an extremely efficient and compelling vehicle to introduce readers to the history, symptoms, and course of Tourette syndrome. For here we have a graphic example of a person, from the nobility no less, whose almost eight decades of suffering seem to provide unique documentation of the course of a disorder and who, by happenstance, was diagnosed and treated by two of France's premier clinicians, Itard and Charcot, more than sixty years apart.

Yet many of the "facts" in the case of the Marquise de Dampierre remain ambiguous and their authenticity depends as much on textual analysis as on medical diagnosis. Most who cite Itard's 1825 article have actually obtained it from Gilles de la Tourette's 1885 report, which is available in English only in an abridged translation.[9] It is fair to say that almost all discussion of the often-cited and emblematic case of Gilles de la Tourette syndrome rests on an 1885 partial reproduction of an 1825 publication in a language inaccessible to many commentators who cite it.

Although Itard's account of the marquise's symptoms is extensive, he examined her only briefly in her twenty-sixth year, eighteen years after the outbreak of her symptoms. Itard's text tells us nothing about the clinical course of the marquise's disorder from 1825 until her death fifty-nine years later in 1884.[10] Despite the fact that the marquise shaped Gilles de la Tourette's classic typology, he never examined or even met her. All of Gilles de la Tourette's clinical discussion about the marquise was appropriated literally from Itard's sixty-year-old text. Finally, contrary to all the implications in the literature since, Charcot never diagnosed, treated, or even talked with the marquise.

The case of the Marquise de Dampierre reveals the extent to which a medical case may be appropriated to legitimate a set of assumptions about the causes and course of a particular set of symptoms and behaviors. Given the way a medical report is written, a reader is led to believe that an objective investigation of a patient's set of symptoms led to the construction of diagnosis and treatment described. An explanation of the cause or etiology of that patient's disorder, attached as it generally is to the conclusion of an article, is presented as if the theory resulted from observation and investigation of the medical case. However, retracing the multiple retellings of the case of the marquise, we discover that the opposite was true. From Itard in 1825, to Gilles de la Tourette in 1885, and to the present, those who selected the Marquise de Dampierre as an exemplar did so to reinforce and to justify an already well-established theory of behavior. Thus, a history of the way in which practitioners have explained the marquise's symptoms tells us as much, if not more, about the history of psychiatry and neurology than it does about the causes of the marquise's suffering. An investigation of the (re)production of the story of the marquise who cursed reveals the power of medical "fictions" in the construction of syndromes. That does not mean this case or the illness is untrue. Rather, it reminds us that emblematic medical cases, like that of the Marquise de Dampierre, and syndromes, like Tourette's, are also literary constructions; and, as literary constructions, they are open to the same sort of textual interpretation and appropriation as any narrative.[11]

The history of Gilles de la Tourette syndrome is both a history of medicine and a history of the production and (re)interpretation of case histories. It does not demean medical researchers to admit this much. No matter how diagnostically informed its classification may be, Tourette syndrome, like the interpretations of texts that have authenticated it, represents historically specific perceptions and expectations of the nature, etiology, and course of behaviors attached to it. A review of the appropriation of the case of the marquise exposes the way medical commentators from 1825 to the present have tried to explain the connection between involuntary cursing and mechanisms of the human brain. Although the terminology has become more sophisticated and both brain science and psychology of the mind have become extremely complicated for a layperson to grasp, the underlying issue that the marquise's behavior exposes remains today, as it was in 1825, one of free will versus biological reductionism.

The marquise's symptoms were the subject of much discussion even before Gilles de la Tourette adopted them as his emblematic case in 1885. The marquise appeared not only in Itard's 1825 article, but also in a number of other texts that, like Itard's, used her symptoms as a vehicle for validating particular medical and philosophical theories of human behavior, as it related to the connection between the functioning of the brain and the existence of the "mind."

Itard's Marquise de Dampierre

Dampierre served as Itard's final example of involuntary movements and speech in a number of patients who seemed to be otherwise in full control of their intellectual functions. Itard described seven men and three women, all of whom displayed bizarre behaviors not unlike those who in earlier centuries were believed to have been enchanted by diabolical forces. Suddenly, seemingly without a precipitating cause, these individuals found themselves compelled to run, jump, scream, or curse, and then, often just as suddenly, the symptoms seemed mysteriously to disappear. It was as if, wrote Itard, "the brain had all at once divested itself of its *authority* on someone's muscular apparatus . . . and that instead of being given up, as one sees in ordinary convulsions, to alternating acceleration and relaxation of contractions, began to execute movements more or less regular, and some functions that they are forced to discharge as if under the exclusive influence of the will."[12]

Itard's understanding of the causes of these behaviors was influenced by the theories of French sensationalist philosopher Étienne de Condillac (1715–1780), who believed that the acquisition of knowledge depended on the stimulation of the senses.[13] As with the Wild Child, Victor, and deaf-mutes in general, Itard believed that those unable to speak from birth had experienced an emotional or social trauma (or irritation) that arrested the physiological development of their brains. But this stunting (or hypoplasia) of cortical development could be overcome, or at least overridden, according to Itard, by careful and persistent education. If Victor's highly developed sense of smell could be understood as resulting from a heightened dependence on his olfactory organs while in the wild, his inability to speak was the result of the underexposure of his brain to human language. Itard believed that the development of language required reinforcement of experience and sensation. The same was true of the human will. A weak will could be strengthened by exercising it. To do this, Itard relied heavily on shaming, very much akin to the so-called moral treatment, then current in insane asylums. Advocates of the moral treatment believed that health depended on the combination of diet, atmosphere, climate, work, and lifestyle; disease, even lesions in the brain, resulted from imbalance of these elements. Fed by a belief that, like the environment, individuals were malleable, suppressive modes of control were de-emphasized in favor of more repressive behaviors. Reflecting wider cultural, economic, and political developments, the moral treatment stressed self-discipline as an alternative to external authority.[14]

Thus, Itard persisted in his attempt to teach Victor to read and speak, and insisted that the mute be taught to talk as well as to learn sign language. Although Itard did allow that some alterations in behavior resulted from irreversible brain damage, he had selected the case of Madame de Dampierre to demonstrate that those like her, who suffered from brain irritations, could benefit from his brand of moral treatment.

In contrast to seven male patients, whom he believed had suffered from irreversible organic brain damage, Itard perceived his three female patients as victims of domestic unhappiness which caused brain irritations. These irritations, in turn, weakened their will, causing the motor tics and involuntary vocalizations.[15] Itard was convinced that his female patients' symptoms were exacerbated or alternatively ameliorated by modifications in their social roles as women, wives, and mothers. In Condillac's terms, as Itard read him, their life experience interfered with their natural development of a female sensibility. Itard's views in this matter meshed with the gendered medical assumptions of his contemporary colleagues who attributed insanity among women to their alienation from the functions prescribed for them within patriarchal culture.[16] Given the proper intervention, this arrested development could be repaired.

Itard connected his first female patient's "sudden involuntary movement," in which she "hit herself in the stomach . . . , striking very vigorously," to the young woman's attempted suicide "after two years of a very unhappy marriage, troubled by sorrows of all types, [and] torments of jealousy." Itard's second female case, the fifteen-year-old Mlle. de C., reacted to unexpected clock chimes and other sounds with "noisy and prolonged outcries," and "loud howling, which one might take from far off for the barking of a dog." Itard tied these symptoms in part to her being an orphan and thus never having been properly trained for her role as a woman.[17]

Itard initially resorted to traditional interventions. The first woman was treated with a blistering agent on her stomach; copious amounts of chicken soup; two three-hour baths; and a two-hour application of a "dozen leeches to the vulva." The second's therapy included baths, general bleeding, applications of cups of leeches, and use of antispasmodics. But in both cases Itard emphasized the moral elements of treatment. It was crucial to keep the first patient separated from "exposing herself too quickly to the influence of the [domestic] causes" of her illness. As for the second, Itard had the young woman taken for walks in Parisian streets, reporting that, due to her fear of public humiliation, "I obtained . . . a success more prompt and more complete than I would have dared to hope. The attacks diminished so rapidly in frequency and intensity, that at the beginning of five weeks Mlle. de C. found herself completely cured."[18]

These two cases formed the context for Itard's final example, that of twenty-six-year-old Madame de Dampierre. When he examined the future marquise, her tics were "continual, not successive and separated by short intervals of a few minutes; sometimes . . . longer, other times shorter, and often even two or three occur unexpectedly following each other without letting up." It was, however, the young woman's involuntary vocalizations that Itard focused on because "they present an extremely rare phenomenon, and constitute an extremely disagreeable inconvenience that deprives the person who is affected of all the sweetness of society."[19]

Although Itard never revealed what he believed were the environmental deprivations that had weakened Madame de Dampierre's will, he asserted that, like his two other female patients, Dampierre's suffering would eventually diminish if she adapted herself to the role of wife and, especially, mother. For, as he explained, although she had married, "the marriage, instead of strengthening and completing her cure, as one would have hoped, very quickly caused her illness to reappear." But that was because "Mme. de D . . . having had no child, was deprived of the favorable benefits that the physical and moral revolution ordinarily provided by maternity would have offered her." Itard lamented that the future marquise was not given over to his care, asserting that "from my observation of" Mlle. de C. "I am persuaded that, if the means of repression or resistance so successfully followed [with her] . . . had been put in service in [Dampierre's case] . . . , we would have obtained the same result."[20]

Itard had highlighted the marquise's symptoms as evidence to confirm his sensationalist claims that otherwise unexplained involuntary behaviors resulted from an underdeveloped will. In Itard's view the will was not localized in any one place in the brain, but like speech or comprehension, it was a function of the entire brain. From that perspective, Itard insisted, in those like the Marquise de Dampierre, who evidenced no localized brain disease, an underdeveloped will was no different from the Wild Child's language deficits. Both could be strengthened by moral treatments.

The Marquise as Metaphor

By midcentury the bizarre behavior of the Marquise de Dampierre was well-known in Parisian aristocratic circles and beyond. A number of physicians,[21] none of whom directly examined or interviewed her, nevertheless cited the marquise's behavior as an example for their theories about the causes of disorders ranging from "illness of the will" to "variable choreas." Although these commentators drew almost all their clinical data about the marquise from Itard's 1825 description, they often used the case to dispute Itard's sensationalism.

In his 1847 three-part study *Maladies de la Volonté,* Ernest Billod cited the Marquise de Dampierre, "well-known in one of the neighborhoods of Paris, afflicted for a number of years from a type of chorea affecting intelligence and the organs of speech," as an example of how a diseased will could lead to an increase of excitation of normally suppressed thoughts.[22] Influenced especially by eclectic philosopher and educator Victor Cousin (1792–1867), Billod, in contrast to Itard, saw the "will" as an organ or "faculty" of the mind that, when diseased, could not be altered by environmental interventions such as reeducation of the senses. Billod rejected Itard's assertion that education and shaming could rid the marquise of her symptoms because, according to Billod, the woman's will was itself damaged.[23] In such cases, wrote Billod, we find "a *lesion* of association of ideas, of memory, of reminiscence, and of imagination." As evidence for his argument, Billod noted that Dampierre's refined (civilized) upbringing and intellect proved insufficient to restrain her subterranean urges for scatological vocalizations and odd public behaviors.[24] His opinion of the difference between normal persons and the marquise was that "we have the power of not expressing all the ideas that come to our mind; the intelligence to effect the choice, and we only express those whose fitness has been appraised." This power "of intellect over expression, is altered in this woman" and as a result her "will . . . submits to other forces" despite what the marquise herself may inwardly desire.[25]

Three years after Billod's articles appeared, David Didier Roth reproduced Itard's description of the Marquise de Dampierre in a book entitled *Histoire de La Musculation Irresistible ou de la Chorée Anormale*. Roth grouped her case with six other published reports, including Itard's Mademoiselle C., in a category he labeled as "muscle tics of speech and the larynx." What united these seven examples, according to Roth, was their inability to control certain sounds, particularly their barking and grunting noises. Rejecting Itard and Billod, Roth argued that vocal tics formed one of the identifiable subsets of "uncontrollable" or "abnormal" choreas, whose underlying cause was physiological (muscular).[26] Roth reinterpreted Itard's description of the marquise's behavior as evidence for an organic etiology of a number of choreic-like symptoms.

The next year (1851) Parisian physician C. M. S. Sandras placed the marquise in the context of Itard's discussion of a man who could not stop walking. Sandras categorized both cases as "chorées partielles."[27] What is interesting is that each physician—Itard, Billod, Roth, and Sandras—read and appropriated the case of the cursing marquise as a metaphor for a particular wider issue, often without reference to the conclusions of previous observers and interpreters. Itard used her as evidence of the difference between cerebral irritations, which could be cured by the moral treatment, versus actual brain lesions, for which other, more traditional medical interventions were appropriate. For Billod, the marquise exemplified an illness of the will. For Roth, she was evidence of the underlying muscular pathology of those choreas that he had differentiated from Sydenham's chorea. And, for Sandras, the marquise served as an example of the variability of chorea.

The final appropriation of the marquise's symptoms prior to Gilles de la Tourette's 1885 publication is found in Théodule Ribot's *Les Maladies de la Volonté* (1883).[28] Ribot (1839–1916), whose book appeared in thirty-two editions in French and English, was professor at the prestigious Collège de France and director of the influential *Revue Philosophique*. After Bénédict Augustin Morel (1809–1873), Ribot was the foremost spokesman for the psychological theory of degeneration in nineteenth-century France. Degeneration theory offered a hereditarian explanation, ultimately tied to moral and social behavior, for a variety of disorders including retardation, depression, depravity, and sterility. Habits including alcoholism, diet, and immorality were alleged to have a cumulative destructive impact on the nervous system that was inherited by succeeding generations. "By adopting a hereditarian approach," writes historian of psychiatry Ian Dowbiggin, "psychiatrists could base their diagnoses on an integrated, biological relationship between body and mind without appearing less medical for recording psychological symptoms."[29] Practitioners took extensive family histories and prepared elaborate pedigrees that sought to explain a current disorder by uncovering patterns of disease and behavior in a patient's family. Although its adherents sought to portray degeneration as organic, treatment revolved around an array of psychological and moral interventions under the rationale that alterations in habits had a direct physiological influence on the nervous system.[30]

Building on the assumptions of degeneration, Ribot separated what he called "impairments of the will" into two categories: The first, reflecting Billod's view, included those who had "an absence rather than an enfeeblement of the will." The second meshed with Itard's belief that in some "the will succumbs or only recovers itself by outside assistance."[31] Citing Billod, Ribot placed the Marquise de Dampierre in the first category of those whose will was diseased: "From the point of view of physiology and of psychology," asserted Ribot, "the human being under these conditions is comparable to an animal which has been decapitated or at least deprived of its cerebral lobes." They reveal "the idiocy of the will or more exactly its madness" and constitute "a permanent pathological state."[32]

Although Gilles de la Tourette appropriated Itard's description of the Marquise de Dampierre as his first and prototypical example, he adopted Ribot's explanations of degeneration for her behavior. "With regard to the innermost *nature* of the affliction, what can be said," asked Gilles de la Tourette at the conclusion of his 1885 essay, "in the absence of any anatomical pathology?" He found his answer in Ribot's *Les Maladies de la Volonté*.[33] Gilles de la Tourette outlined an inevitable path from involuntary motor and vocal tics in early childhood that led eventually to cursing.[34] There was no hope, wrote Gilles de la Tourette, of "a complete cure"; for "once a ticcer, always a ticcer."[35] This assertion was not based on Gilles de la Tourette's clinical experience, but rather it came from his reading of the case of the Marquise de Dampierre, a woman neither he nor Charcot ever examined.

The claim that "the famous Marquise [was] a patient of Charcot,"[36] rests no doubt on Gilles de la Tourette's 1885 article: "In 1825," he begins, "Itard published a study . . . that . . . is extremely conclusive and even more interesting since the sick woman who is in fact our subject lived until 1884 and had been seen by professor Charcot, who had retrospectively verified the diagnosis."[37] A few lines later he reports that "Professor Charcot *saw* this *sick person* frequently, who until advanced age, still manifested her incoordination and continued, despite herself, to say obscene words that Professor Charcot witnessed, even in public places."[38]

In fact, Charcot literally only *saw*, that is witnessed, the marquise's behavior. Charcot mentioned the marquise five times in his published lectures. As most of us know from personal experience, lecturers often repeat the same story slightly differently over time, and Charcot's explication of his encounter with the Marquise de Dampierre was no exception. What is unambiguous in four of these lectures, however, is that Charcot never had any direct contact with the marquise, let alone any physician-patient relationship. In

Figure 2.2 Georges Gilles de la Tourette, neurologist at the Salpêtrière Hospital in Paris, published an article in 1885 describing a combination of involuntary motor movements and spontaneous shouting and cursing as a single disease that he called "maladie des tics" (known today as Tourette syndrome).

December 1887 Charcot reported that "there was in high Parisian society, someone of the most aristocratic part of society, who was known for uttering filthy words. *I never had the honor of knowing her;* I did recognize her one day climbing the stairway to the Salon and I was surprised to hear her suddenly say SNdD [Holy name of God]."[39] Only in a report published in Italian in 1885 by the physician Giulio Melotti (of Bologna), who had spent 1884 and 1885 in Paris attending Charcot's lectures, is Charcot alleged to "often had the opportunity to examine [*esaminare*]" the marquise.[40] But, Melotti either assumed that Gilles de la Tourette's statements that Charcot had often observed the marquise's behavior were equivalent to a medical examination or Melotti's choice of the verb *esaminare* was due to a mistranslation of the French verb *voir* [to see] in Gilles de la Tourette's report.[41] What is most revealing, however, is that Melotti's rendition adds no new information about the marquise's symptoms that was not already available in Itard's 1825 report.

Gilles de la Tourette's assertion that "Professor Charcot saw this sick person frequently," was most likely a rhetorical device whose aim was to justify inclusion of an example more congruent with Gilles de la Tourette's typology than his actual clinical observations. Suggesting, albeit circumspectly and ambiguously, that the marquise was treated by Charcot lent clinical legitimacy to an example that otherwise would have been merely anecdotal. Despite the numerous citations of the case of the Marquise de Dampierre, a careful reader will quickly discover that every clinical statement made about her can be found in Itard's 1825 account. The evidence that her cursing persisted until her death came from obituaries, rather than from any subsequent clinical examination.[42]

"Study of a Nervous Affliction" 1885

Charcot, who customarily used his students to float his new ideas,[43] chose Gilles de la Tourette because of the younger man's clinical experience with ticcing patients and because of his wider interests in similar phenomena.[44] In particular, Gilles de la Tourette had been fascinated by the relationship between tic disorders and jumping and startle behaviors reported in Malaysia, Siberia, and Maine. These bizarre behaviors, like TS, often were accompanied by imitative behaviors, mimicry, involuntary cursing, and sexually explicit gestures and displays.

The most influential and detailed study was American neurologist George M. Beard's description of the startle, jumping, and tic-like behaviors of a number of French Canadian lumberjacks living in Maine. Beard, who popularized the diagnosis of neurasthenia, a condition he asserted was caused by exhausted nerves, visited and "experimented" with these lumberjacks near Moosehead Lake. He published his findings in an 1880 report to the American Neurological Association, and in slightly different form, in the widely circulated journal *Popular Science Monthly*.[45] Beard was struck particularly by these men's suggestible and imitative behaviors. "One of the jumpers," reported Beard, "while sitting in a chair with a knife in his hand was told to throw it, and he threw it quickly; . . . at the same time he repeated the order to throw it, with a cry or utterance of alarm resembling that of hysteria or epilepsy."[46] Two others "were told to strike, and they struck each other very forcefully." Another "was suddenly commanded by a person on the other side of the window, to jump, and he jumped straight up half a foot from the floor, repeating the order." Beard reported that when one of the lumberjacks was read lines in English that he did not understand, the man, nevertheless, "repeated or echoed the sound of the word as it came to him, in a quick, sharp voice, at the same time he jumped, or struck, or threw, or raised his shoulders, or made some other violent muscular motion." Beard found that these men "could not help repeating the word or sounds that came from the person that ordered them any more than they could help striking, dropping, throwing, jumping or starting." He called all of these behaviors "jumping" and was convinced that they were hereditary.[47]

Gilles de la Tourette translated and published Beard's observations in 1881.[48] Three years later he published a second article comparing jumping with similar phenomena reported in Malaysia and Siberia called "latah" and "myriachit."[49] Gilles de la Tourette drew a parallel between these reports and similar cases that had recently presented themselves at Charcot's clinic, including that of a fifteen-year-old boy who was "afflicted with an extreme hyperexcitability, with particular tics, convulsive movements of his head and trunk, after which he almost invariably shouts out the word of Cambronne [shit]. Moreover, if someone happens to speak in front of him, he faithfully repeats the two or three words which ended the sentence."[50] The next year when Gilles de la Tourette published his two-part "Study of a Nervous Affliction," he included jumping, myriachit, and latah as variations of his typology.[51]

Gilles de la Tourette's article laid out a typology of symptoms and prognosis based on his description of nine patients' case histories and three examples of latah, myriachit, and jumping. Convulsive tic illness, according to Gilles de la Tourette, was a chronic disease that developed from childhood motor and vocal tics ultimately to cursing.[52] Unlike other choreas and hysterias, "la maladie des tics" was both progressive and hereditary; it might wax and wane, but ultimately resisted all interventions.[53] Yet none of Gilles de la Tourette's actual clinical observations sustained such assertions.

In fact, the oldest patient Gilles de la Tourette had treated when he published his landmark 1885 study was Ch., a twenty-four-year-old clerk, who at eight years developed facial twitches, followed by arm and leg movements. Ch. displayed knee bends, lip biting, and jumping; however, he evidenced no involuntary sounds or coprolalia. The second oldest of Gilles de la Tourette's patients, M. E., was a twenty-one-year-old draftsman who first exhibited eye blinking at age eight or nine and then contortions that led to a changed style of walking. M. E. also jumped if brushed accidentally, but he too displayed no vocalization abnormalities.[54]

The third oldest, S. J., a twenty-year-old civil servant, displayed an array of symptoms, including motor and vocal tics, cursing, word repetition, and jumping. However, Gilles de la Tourette reported, S. J.'s cursing unaccountably ceased and his movement disorders, except for some eye tics and infrequent tongue protrusions, disappeared. Only some echolalia (word repetition) remained. Gilles de la Tourette's three other clinical cases were eleven-, fourteen-, and fifteen-year-old patients. Like the first three cases, none of them were old enough to sustain Gilles de la Tourette's or Charcot's assertions that the syndrome was a lifelong or progressive affliction. Fourteen-year-old Ch. (not to be confused with the older Ch.) developed face twitches at age eight, followed by involuntary arm, leg, and right-side movements. "The child also made bizarre contortions, bending his knees [and] jumping in place," but neither made involuntary sounds nor cursed. Sometimes Ch. "opened and closed his mouth with so much force that his bottom lip bled."

But a year later, reported Gilles de la Tourette, all these movements "have extremely diminished, especially in intensity." J. L., eleven years old, began to grimace at age seven and soon developed general muscular twitches. Although he displayed no involuntary cursing, he often uttered loud cries of "ouh! ouh!" Fifteen-year-old G. D. first experienced motor tics when he was eight, which ameliorated for four years. Symptoms returned at age twelve, this time accompanied by coprolalia, during which G. D. would shout out "shitty asshole."[55]

Thus, only two of Gilles de la Tourette's clinical observations displayed the range of symptoms that he and Charcot attached to the general syndrome. None, however, fit their assertions that the syndrome was unambiguously progressive and lifelong. Indeed, the twenty-year-old civil servant, S. J., who had exhibited florid symptoms, was completely free of them (except some residual word repetition) the year after Gilles de la Tourette examined him. In this context, then, the example of the Marquise de Dampierre was essential as evidence for Gilles de la Tourette's claim that he had described a syndrome that must be distinguished from other seemingly similar disorders.

Those who cite the case of the marquise in current clinical literature have continued to elaborate on her symptoms, authenticating the notion that hers was a carefully observed clinical case whose long course provided the bedrock for both diagnosis and prognosis of Gilles de la Tourette syndrome. This is significant because unlike Gilles de la Tourette's own clinical cases, only the Marquise de Dampierre unambiguously qualifies as suffering from Gilles de la Tourette's disease as originally described and Tourette syndrome as it is now represented in the DSM-IV. In contrast, the symptoms and course of the patients Gilles de la Tourette had actually observed were much more ambiguous.[56] As a result, none have ever become embedded in the clinical history of the syndrome. Without the Marquise de Dampierre, Gilles de la Tourette would not have been able to provide an example that unequivocally sustained his clinical claims.

The issue is *not* whether the Marquise de Dampierre can be viewed legitimately as having suffered from what we now call Gilles de la Tourette syndrome. Clearly, the marquise suffered from a neurological deficit that, for want of an identified underlying pathology, is best described as Tourette syndrome. But to the extent that her life history has become an emblem of the affliction, not only have certain symptoms been privileged, but also others have been ignored or downplayed. In raising the marquise's symptoms to an emblematic level, those with less florid or more ambiguous symptoms often were denied a diagnosis. Indeed, it was partly the rigidity of Gilles de la Tourette's classification that undermined its diagnostic usefulness by the late nineteenth century.

This has had two important consequences, one historical and the other clinical; although the boundaries separating medical history from clinical practice are less demarcated than is often supposed. In essentializing Gilles de la Tourette's particular reading of Itard's text, other possible meanings of Dampierre's symptoms have been lost. Those who wish to trace the history of tic illnesses have searched for clinical cases that mesh with the symptom course that Gilles de la Tourette ascribed to the marquise.

As important, the construction of emblems has a profound influence on clinical practice and medical research because all claims about the epidemiology of a disorder (for instance, its increase or decrease over time) depend finally on a unitary definition. Because the ultimate purpose of emblematic cases is to alert and assist practitioners in diagnosis and treatment, these narratives retain a profound influence on those afflicted with elusive symptoms like those associated with Tourette's.

As long as the history of these symptoms is set within the framework of Charcot and Gilles de la Tourette—that is, the story of the marquise's symptoms—the only possible conclusions are that Gilles de la Tourette accurately defined the boundaries of this illness or that he was in error about a number of its elements. Yet many of his contemporaries refuted Gilles de la Tourette's typology. An alternative perspective, not based on the case of the marquise, reveals a different history with an enlarged cast of players and another medical focus. In such a history, the story of the marquise is contextualized as just one among many possible descriptions, leading us away from Gilles de la Tourette's description of a separate disease toward a continuum with similar movement disorders.

A Disputed Illness

...3...

Late twentieth-century medicine has adopted Gilles de la Tourette's description of symptoms as the designation for a clinical syndrome. Simultaneously, it has ignored his and Charcot's insistence that "maladie des tics" was a progressive illness, in which the afflicted inevitably developed involuntary cursing (coprolalia).[1] The disease might wax and wane, but ultimately resisted all interventions.[2] Even less well known today is Gilles de la Tourette and Charcot's belief that they had identified a hereditary (degenerative) cause of these symptoms.[3]

Because few ticcing and cursing patients seemed to display either the degenerative family history or the inevitable symptom course that Gilles de la Tourette had described, most physicians were not persuaded that the symptoms of "la maladie des tics de Gilles de la Tourette" constituted a separate disorder. Focusing on the issues of symptoms and course of illness, many of Charcot's colleagues and students were convinced that Gilles de la Tourette's classification was, instead, a florid manifestation of hysteria. Outside this Salpêtrière circle, most others believed that multiple motor tics and coprolalia were a subset or variation of movement disorders known as choreas, which were widely regarded as having a common cause in a previous attack of rheumatic fever. As a result of these two strands of criticism, "la maladie des tics de Gilles de la Tourette" was increasingly marginalized. By the first decade of the twentieth century, Gilles de la Tourette's disease had virtually disappeared from the medical map.

Figure 3.1 In 1885 Jean-Martin Charcot (chief physician of the Salpêtrière Hospital) directed his clinical chief, Georges Gilles de la Tourette, to collect and publish cases of involuntary vocalizations and motor movements combined with cursing. Charcot renamed the disorder "maladie des tics de Gilles de la Tourette" in honor of his intern.

These two sets of attacks on the need for a separate disease classification called maladie des tics were parallel rather than integrated, and their stories are best told separately. The first, coming from within Charcot's own Salpêtrière, would prove to be the most damaging in the short run. The second, suggesting a rheumatic etiology for Tourette's, would turn out to be the most sustained and, ultimately, the most resilient.

A Form of Hysteria

Charcot's classification of "la maladie des tics de Gilles de la Tourette" met with almost immediate rejection from a number of influential clinicians, including some of his most loyal followers. A general objection to Gilles de la Tourette's typology came from Charcot's contemporary and sometimes collaborator, Valentine Magnan (1835–1916), at l'Hôpital Sainte-Anne.[4] Magnan protested that "tics, echolalia, coprolalia, and the other bizarre phenomena" that Gilles de la Tourette "had bundled together, did not form a true illness," but rather were merely forms of "mental degeneration."[5]

The most persistent and ultimately most effective criticisms came from Salpêtrière neurologist Georges Guinon, who published three essays in 1886 and 1887 contesting Gilles de la Tourette's diagnostic conclusions.[6] Focusing on the course of ticcing and cursing patients' illnesses, Guinon argued there was no need to erect a separate category of diagnosis called maladie des tics. Like Magnan, Guinon believed that these symptoms were a continuum because of their common etiology in hereditary degeneration.[7] Guinon also claimed that other symptoms, including what later would be called compulsive behaviors, had been ignored by Gilles de la Tourette.[8] Most important, drawing his evidence from Charcot's cases and lectures, Guinon pointed out that although all hysterics displayed some involuntary movements, severe cases also exhibited the most florid symptoms of convulsive tic disorder.

Charcot had been instrumental in defining hysteria as that class of disorders in which patients displayed somatic symptoms, like paralysis, for which no underlying somatic cause could be found. According to Charcot, hysterics had experienced an earlier psychological trauma, for instance, a railway accident, but had emerged physically unharmed. Subsequently, they developed paralysis, which served as a sort of protection from placing themselves in future dangerous situations.[9] By demonstrating that these paralytic symptoms disappeared under hypnosis, Charcot believed that he had proven that the disorders were hysterical rather than neurological. But Charcot's view of hysteria was not entirely psychological; he also insisted that hysteria was an inherited disorder and that only those who were truly hysterical could be hypnotized.[10]

As Guinon had pointed out, Charcot often had described the same ticcing and vocalization behaviors in hysterics that Gilles de la Tourette had attached to tic illness. Charcot, however, asserted that hysterical tics resulted from imitation and could be cured "easily" through "appropriate means," such as hypnosis, while the tics described by Gilles de la Tourette were inherited and resisted all treatment.[11] Thus, when tics and vocalizations persisted, resisting all interventions, a differential diagnosis of "maladie des tics de Gilles de la Tourette" was appropriate. In these cases, Charcot and Gilles de la Tourette asserted that the underlying pathology could be identified in the hereditary history of their patients. Charcot's definition of predisposing family pathologies, including nervous habits, alcoholism, and epilepsy, was promiscuous. He diagnosed a twenty-one-year-old woman with tic illness because "her mother was very nervous, subject to unbridled anger: several times in a fit of jealousy, she attempted to poison herself; her maternal grandfather . . . died of grief following a reverse of fortune; she has a brother subject to great fits of anger."[12] Based on this and similar family histories, Charcot insisted that hysteria and maladie des tics were separate disorders: "hysteria and the maladie des tics," Charcot explained in March 1886, "can coexist, but the latter cannot be derived from the former or conversely."[13]

However, Guinon insisted that careful observation over time demonstrated that sufferers did not fit Gilles de la Tourette's assertion of a life-long progressive disorder. Citing his patient, sixty-four-year-old Mme J., he reported that "to the contrary, outside her periods of illness," Mme J. was completely normal. She exhibited "not the tiniest grimace, not the smallest convulsive tic of her limbs" during her latent intervals.[14] The case of a thirty-eight-year-old railway employee who displayed tic disorders and compulsive behaviors, including the need to count to seven whenever he got up in the morning, convinced Guinon that convulsive tic disorders were often informed by what today might be characterized as obsessive-compulsive behavior. Two other of Guinon's patients exhibited symptoms—involuntary movements, coprolalia, and echolalia—that were consistent with those outlined by Gilles de la Tourette, but also both imitated movements they observed in others. Adopting Charcot's label of this echo behavior as "echokinésie," Guinon argued that these patients acted similarly to patients who were hypnotized; yet, according to Charcotian theory, only hysterics could be hypnotized. If convulsive tic disorder was a variation of hysteria, then Gilles de la Tourette's typology was based on an incomplete set of clinical observations.[15]

Guinon challenged Charcot's assertion that hysteria and maladie des tics could coexist, but were unrelated. Reproducing three cases, Guinon showed that in the most severe cases these "two afflictions, hysteria and convulsive tic illness, often can be found associated with one another."[16] Although all hysterics displayed some involuntary movements, severe cases, argued Guinon, also exhibited the most florid symptoms of convulsive tic disorder. Guinon concluded that convulsive ticcing behavior, rather than being a separate disorder, was merely a florid form of hysteria.[17]

Charcot's initial reaction to Guinon's criticisms was to downplay the dispute by extending credit to both his students, while also emphasizing the importance of viewing Gilles de la Tourette's typology as distinct from hysteria. Using the case of a thirteen-year-old ticcer with coprolalia, Charcot credited both Gilles de la Tourette and Guinon as having worked extensively with patients who suffered from maladie des tics.[18] Nevertheless, Charcot rejected Guinon's claim that convulsive ticcing behavior was a form of hysteria by insisting that maladie des tics was a combination of inherited psychological and physiological causes: "Every time that you see someone afflicted with maladie des tics, if sometimes only slightly, but especially when it is accompanied with phenomena like those of coprolalia, you are certain of encountering the direct product of hereditary insanity."[19] According to Charcot, what separated maladie des tics from hysteria was not its symptoms, but its etiology.[20] He contrasted the case of a twenty-two-year-old woman whose tics and barking noises seemed similar to that of a seventeen-year-old boy who he had diagnosed as afflicted with maladie des tics. Charcot claimed that the woman suffered from hysterical tics because her barking could be controlled by hypnosis, while the young man's could not. Thus, Charcot argued, "the difference between barking ticcers and barking hysterics" is that "hysterical barking can be cured."[21]

The fact that Charcot repeated these arguments so frequently over the next two years suggests that he was reacting to great resistance among his colleagues and students in acknowledging that Gilles de la Tourette's disease was a distinct disorder.[22] His final attempt to differentiate hysterical tics from Gilles de la Tourette's disease appeared in an 1890 doctoral thesis he directed, written by Jacques Catrou.[23] Gilles de la Tourette supplied Catrou with twenty-eight of his previously unpublished cases.[24] Charcot (and Gilles de la Tourette), through Catrou, admitted that hysterics often displayed the range of symptoms that had been attached to Gilles de la Tourette's illness—including the most florid of these, cursing and echolalia. However, using a case supplied by Gilles de la Tourette, Catrou argued that Charcot had demonstrated that in hysterics these behaviors were always imitative because hysterics easily succumbed to suggestion (thus, only hysterics could be hypnotized).[25] But Catrou claimed that in "la maladie des tics de Gilles de la Tourette" these behaviors were always spontaneous and never imitative. Moreover, Catrou explained that hysterics always could be cured of their tics either by hypnosis or other suggestion. True sufferers of Gilles de la Tourette's illness, on the other hand, faced a much worse prognosis. Although these patients' symptoms might wax and wane, the involuntary movements always returned and seemed to evolve from milder to more extreme manifestations, in which involuntary cursing appeared.[26] What distinguished Gilles de la Tourette's illness from a hysterical illness with similar symptoms was whether or not the cause was hereditary and whether or not the symptoms could be eradicated.

By 1893, the year of Charcot's death, the issue of what constituted "la maladie des tics de Gilles de la Tourette" was becoming even less clear. Julien Noir, a student of one of Charcot's most loyal supporters, Désiré Magloire Bourneville of the Bicêtre Hospital in Paris, reproduced seventy-five Bicêtre cases, including extensive photographs of patients with tics, coprolalia, echolalia, echokinesia (need to imitate), compulsions, and their combinations. Offering a comprehensive review of the debates over "la maladie des tics de Gilles de la Tourette," Noir distinguished four varieties of tic disorders.

Noir called his first type "tic convulsifs simples." These were "purely motor phenomena," whose antecedents could be found in a reflex reaction to an earlier irritation that these suggestive persons continued to perform long after the source of irritation had dissipated. While many persons experienced irritation, only those who inherited nervous dispositions were especially prone to simple convulsive tics.[27] The second type, "coordinated tics," were found in idiots and other mentally retarded individuals. These behaviors meshed with those tic patients Magnan and his students had described as mental degenerates.[28] The third type, "maladie des tics psychiques," were "purely psychological phenomena." Psychic tics, according to Noir, were the most common form of convulsive tics. They could include both compulsive behaviors as well as the more florid cursing and echo behaviors. Among psychic convulsive tics, Noir included inappropriate vocal outbursts, as well as latah, myriachit, and jumping.[29] Finally, Noir explained, when "echolalia, coprolalia, [and] echokinesia" were "added to simple convulsive tics, they form the syndrome described by Gilles de la Tourette which deserves, because of it evolution, to be kept in a separate part of pathology." This illness "separates itself completely from other tics," because its symptoms worsen as they evolve, with more florid symptoms replacing less complex ones.[30] Noir's thesis raised as many questions as it answered, further circumscribing and marginalizing Gilles de la Tourette's illness by validating the observations of his critics that the majority of motor and vocal tics were not maladie des tics.

Almost immediately after Charcot's death, other former members of Charcot's circle endorsed criticisms of maladie des tics.[31] Most important among these critics was Édouard Brissaud who was temporarily appointed to fill Charcot's professorship in 1893.[32] In a series of articles based on his clinical cases at l'Hôpital Saint-Antoine, Brissaud labeled the irregular motor movements, sometimes accompanied by coprolalia and echolalia, as "chorée variable" because its most distinguishing feature was always the irregularity and unpredictability of its motor symptoms, as opposed to the regularity of motor movements found in Sydenham's and Huntington's choreas.[33] Unlike a specific inherited cause as in Huntington's, variable chorea was attached to degeneration.[34]

In 1899 Gilles de la Tourette, now Professeur Agrégé in Legal Medicine at l'Hôpital Saint-Antoine, published an article in *La Semaine Médicale,* responding directly to the claims made by Brissaud and Guinon.[35] Although he preferred the term "unbalanced" to "degenerate," Gilles de la Tourette accepted Guinon's insistence that the syndrome "almost always exposed a condition of mental instability characterized by numerous phobias, of arithmania [compulsive counting], agoraphobia [fear of crowds], and all stigmata which today are referred to as mental degeneration." But Gilles de la Tourette was unwilling to concede Guinon's major criticism—that convulsive tic disorder was merely the most extreme manifestation of hysteria.[36] In his response to Brissaud, he argued that "an affliction that M. Brissaud described in 1897, under the name of *variable choreas of degenerates,*" was "not to be differentiated from the illness of convulsive tics," and that Brissaud's patient actually fit Gilles de la Tourette's typology.[37]

A Form of Chorea

If critics from within Charcot's circle were unrelenting in their rejection of Gilles de la Tourette's typology, they at least shared Charcot's belief that these ticcing and cursing patients suffered from a psychological disorder enabled by a familial inheritance. However, an entirely different set of explanations for these bizarre symptoms predated Gilles de la Tourette and persisted long after he described his separate illness. The long-held belief was that the involuntary movements and vocalizations described by Gilles de la Tourette were a subset of choreas, in which the afflicted present a ceaseless occurrence of a wide variety of rapid, jerky, involuntary, but well-coordinated movements.

Modern textbooks delineate the symptoms of chorea's most common form, Sydenham's, as including facial movements such as frowning, raising the eyebrows, pursing the lips, smiling, and bizarre, generally bilateral, movements of the mouth and tongue.[38] Long called "Saint Vitus's dance," chorea first was fully described in 1686 by British physician Thomas Sydenham (1624–1689), who selected the Greek term *Khoreia* (for dance), and labeled St. Vitus's dance as "chorea minor."[39] By the end of the nineteenth century the designation "Sydenham's chorea" was adopted for the most common manifestation of choreic movement disorders. However, throughout the nineteenth century the term "chorea" was widely rather than narrowly proscribed. It was used by physicians to describe a number of highly variable movement disorders that included Sydenham's chorea, tics, dystonia (muscle tone impairment), myoclonus (muscle spasms), as well as those behaviors called "convulsive tics" that Charcot labeled as "the tic disease of Gilles de la Tourette."[40] Although medical commentators throughout the nineteenth century observed clusters of symptomatic differences that might justify separating each form of chorea as a distinct disorder, they were reluctant to do so because of a widespread belief that most, if not all, choreas were connected with a prior attack of rheumatic fever.[41]

The observation that rheumatic fever was a crucial predisposing feature of choreic movement disorders was first made by a number of late eighteenth- and early nineteenth-century European physicians. As early as 1802 the *Syllabus, or Outlines of Lectures on the Practice of Medicine*, published by London's Guy Hospital, concluded that rheumatism was "one of the existing causes of chorea" because so many choreics had reported an earlier attack of rheumatic fever.[42] Across the Channel French physicians reached parallel conclusions.[43]

By the nineteenth century, based on autopsies that revealed heart damage in Sydenham's patients and the frequent coincidence of subsequent involuntary tics and motor movements following rheumatic illness, the vast majority of medical commentators were convinced that there was a causal connection between rheumatic fever and Sydenham's chorea. Citing numerous published clinical reports of British physicians, the distinguished British physician Richard Bright could write in 1838 that the "combination of . . . spasmodic disease with rheumatism has long been recognized."[44] In 1850 French physician J. P. Botrel, drawing on the work of English physicians, wrote a dissertation entitled "Of Chorea Considered as a Rheumatic Affliction," whose thesis was that chorea was one of a number of possible manifestations of rheumatism.[45] That same year the French Academy of Medicine published and awarded a prize to Germain Sée of the Hôpital des Enfants for his *De la Chorée,* which concluded that "in most cases . . . chorea results from a rheumatic predisposition."[46] Although Sée had reached this conclusion from his own clinical experience, he noted that he had discovered "numerous examples of rheumatic chorea in medical journals," of which he listed more than thirty published in Britain and France between 1810 and 1845.[47] Thus, by 1862, Armand Trousseau could write with authority that "of all the predisposing pathological states [of choreas], rheumatism is assuredly the most marked and the least questionable."[48]

Tied to the view that rheumatic disorders often, but not always, resulted in choreic movement disorders was the belief that variations in movement symptoms were different manifestations of a common underlying condition, not distinct diseases. This made sense because of a great variety in the severity of symptoms and outcome among those afflicted with rheumatic fever: some patients died, others recovered, but with weakened heart function, and a number seemed to display no residual symptoms. It was logical to expect that the involuntary movements that followed a rheumatic illness would also vary in form, severity, and outcome, perhaps due to the nature of the previous rheumatic affliction or to the different constitutions of the afflicted. Thus, Bright insisted in 1838 that rheumatic fevers could cause an array of different disorders depending on which organs of the body had been diseased.[49] This view was elaborated in Paris in 1850 by Sée, who categorized choreas into a variety of "pathological states" ranging from their principal manifestations as "irresistible and uniform movements," to a number of "anomalous choreas," including those forms whose sufferers displayed symptoms that today would be identified as Tourette syndrome. No matter what the particular pathological manifestation, Sée found that clinical evidence demonstrated that all "these diverse choreas" were ultimately connected by the predisposing cause of "rheumatism."[50]

Perhaps the clearest midcentury statement of the view that a predisposing rheumatic cause undercut the need for construction of separate syndromes for each variety of motor movements was made in 1851 by C. M. S. Sandras, of the Faculty of Medicine of Paris. Sandras identified two basic types of chorea, acute *(aiguë)* and chronic *(chronique)*.[51] Acute chorea, according to Sandras, was less common, yet more progressive and lethal than chronic chorea, which manifested itself either as general or partial (convulsive). General chronic chorea fit Sydenham's chorea, while Sandras's examples of partial or convulsive chorea (which he also called "convulsive tics") meshed with those involuntary tics and vocalizations that would become classified later as Gilles de la Tourette's tic disease. Sandras's primary example of partial chorea was Itard's case of Madame de Dampierre.[52] Although he connected the etiology of both acute and chronic choreas with rheumatism, chronic choreas, according Sandras, "appear [more] often with or after genuine attacks of rheumatism."[53] Nevertheless, Sandras resisted asserting that his classifications constituted three or even two *distinct* diseases with *distinct* causes. Rather, Sandras pointed out (as others later would) a major drawback of attempting to construct separate syndromes based on differing symptomatic manifestations: frequently, explained Sandras, those suffering a seemingly discrete form of chorea would develop the symptoms attached to another form. "Acute chorea is often replaced," found Sandras, "by a chronic [general] chorea or by convulsive tics." In fact, Sandras wrote, it was often extremely difficult to distinguish among these varieties because those "chronic [general] choreas which replace acute chorea are those which show the greatest tendency to revert to one of the more serious forms of the illness."[54]

Lacking any firm neuropathological evidence or understanding of the mechanisms of infections in the brain, nineteenth-century practitioners interested in understanding and treating choreas continued to fall back on nosological approaches (classification by symptoms) as one of the few tools in their medical armamentarium. Trousseau's influential two-volume *Clinique Médicale de l'Hôtel-Dieu de Paris* (republished in eleven editions in France, Great Britain, the United States, and Germany from 1861 to 1913) placed those symptoms on which Gilles de la Tourette would base his disease into two separate categories, "spasmodic" (motor) tics and "laryngeal" or "diaphragmatic" (vocal) tics.[55] What distinguished these from Sydenham's chorea were not their onset or uncertain course, but their frequency and the vocalizations. Having made this distinction, Trousseau produced a case of a fourteen-year-old boy "afflicted with extremely severe tic" *and* uncontrolled vocalizations, calling into question the very boundaries that he had erected. Even if choreas ought to be separated by symptoms, Trousseau, nevertheless, insisted that the *"hereditary predisposition"* of all choreas was "unquestionable."[56] Thus, in terms of predisposing causes, course of illness, and symptomatic boundaries, Trousseau had actually provided as much evidence for the proposition that all choreas were variations of the same disorder as the conclusion that each variety ought to be seen as a unique illness.

Throughout the nineteenth century the vast majority of physicians who tried to understand choreic movement disorders focused on attempts to link symptoms and etiology. Assuming that variations of choreas shared a common rheumatic fever pathology, these practitioners resisted creating a separate disease category for convulsive tics. The great and extremely influential exception to this rule was Charcot.[57] In conflict with more than eighty years of observations by the most well-regarded physicians in France and Britain, Charcot rejected any connection between rheumatic fever and choreas.[58] It was his view that "because these two afflictions are part of the same pathological family *(la 'famille arthritique'),* . . . individuals who are affected by one of these illnesses are often more susceptible to having the other."[59] Charcot insisted that symptomatic differences and the course of each illness required separating variable choreas [convulsive tics] from Sydenham's chorea.

Following this view, modern clinicians have differentiated their diagnoses of what has become known as Gilles de la Tourette syndrome from Sydenham's chorea and other movement disorders by emphasizing *symptomatic* differences. That is, more regular and coordinated movements are seen as Sydenham's in contrast with the irregular and uncoordinated tics associated with Tourette's. What has gone largely unnoticed is that Charcot had described the underlying hereditary causes of Gilles de la Tourette's illness and Sydenham's chorea as virtually indistinguishable from one another. He had drawn on Théodule Ribot's theory of hereditary psychological degeneration to explain the etiology of convulsive tics.[60]

"Heredity," wrote one of his students, was the "predisposing cause of all tics," while "the defect in inhibition of the will" was "the indispensable condition for their reproduction."[61] Yet Charcot also asserted that "the real cause of [Sydenham's] chorea resided in hereditary transmission."[62] If one searched the family trees of those with Sydenham's chorea, explained Charcot, one might discover on one branch familial "antecedents" including "gout, rheumatic fever, certain types of migraines," and enteric diseases, while on the other branch there might be "neurasthenia, hysteria, epilepsy," and other hereditary disorders such as "general progressive paralysis" and "locomotor ataxia."[63]

Because Charcot offered almost identical lists of the hereditary antecedents of maladie des tics and those of Sydenham's, most physicians outside the Salpêtrière circle were persuaded motor tics and cursing shared a common etiology with Sydenham's chorea. And, contrary to Charcot's assertions, most of these physicians continued to assume a connection between prior rheumatic disease and subsequent tic behaviors.[64] In 1891 E. Cadet de Gassicourt of the Hôpital des Infants and a member of the Académie de Médecine specifically rejected Charcot's assertions that there was no connection between "rheumatism" and tic disorders. Although "the rheumatic . . . etiological connection [with chorea] has never been accepted without contest," wrote Cadet de Gassicourt, "the work of M. Germain Sée, more than forty years ago" as well as "the results of my, unhappily extremely long, practice have left me no doubt, not only on the coincidence between rheumatism and chorea, but also on the identical nature of one another."[65]

By the 1890s the connection between rheumatic fever and chorea was so well established that one is hard-pressed to find much dissent in North American and European (including French) medical journals. Writing in the *Montreal Medical Journal* in 1891, neurologist George Brown reviewed his own patients' records, medical journal reports, and statistics of hundreds of other cases and concluded that "in the majority of cases it [chorea] is a rheumatic manifestation."[66] Detroit physician W. M. Donald noted in 1892 that among physicians there was widespread agreement that rheumatism and chorea, in terms of "cause and effect stand so closely interrelated, that no room for doubt is left in the mind of even the most superficial observer . . . so that on a case of chorea presenting itself to a physician the question as to a pre-existence of rheumatism should naturally arise in the mind."[67] Even those who argued that rheumatic fever was only one possible cause of chorea never completely dissociated the two. For instance, Dr. Vincent Pagliano of Paris suggested that similar symptoms could have a variety of underlying pathologies brought on by a number of different "infectious illnesses." Nevertheless, Pagliano cited "the words of M. Cadet de Gassicourt, which seem to me to be an exact expression of truth: All choreas are not rheumatic; but it would be an exaggeration in the reverse to deny the action of rheumatic fever in a certain number of cases."[68]

Because late twentieth-century Western medicine has adopted Charcot's typology of movement disorders, historians of medicine and clinical practitioners have neglected the late nineteenth-century medical discussion about the possible infectious substrate of these disorders. Symposia and review articles on the possible connection between rheumatic fever and subsequent movement disorders were reported and published in local, regional, and national medical journals in the United States, Canada, Britain, and France. Some of the most important names in North American medicine participated in these discussions, including Weir Mitchell, M. Allen Starr, and William Osler.[69]

The new science of bacteriology spurred on this intense interest. As early as 1883 a Virginia physician reported that one of his patient's symptoms supported the view of the "the *unity* of origin of acute rheumatism, carditis, and chorea—that origin being the pathological properties existing in the white blood corpuscles as the result of *follicular tonsillitis*."[70] In 1891, a Parisian physician argued that the underlying cause of rheumatic fever followed by chorea was a "microbial blood infection."[71] By 1892, physician Henry P. Cooke, citing the works of the Germans Laufenauer and Koch, reported that the same "virus, affecting the serous surfaces may produce tissue changes and a chain of clinical phenomena constituting rheumatism; and the same virus affecting certain areas or tracts in the nerve centers, either directly as an irritant or indirectly through the agency of reflexes, or through products of its action on other tissues, may originate the irregular muscular movements and associate disturbances . . . which are designated chorea."[72] Although Cooke used the term "virus," the distinction in 1892 between viral and bacteriological agents was not yet firmly made. In any case, Cooke's evidence actually pointed to bacterial infection.

One of the most remarkable studies was published by an Italian neurologist, Robert Massalongo, in the French *Revue Neurologique* in 1895. Massalongo, like many of his contemporaries, had rejected distinct classifications for movement disorders. "Chorea, let us not forget," wrote Massalongo, "is not one morbid entity, but a *clinical syndrome;* etiologically there are many choreas, as there are many epilepsies, but in the last analysis it is a question of transitory or permanent irritation of the neural motor systems by toxic substances, of a self-poisoning ["auto-intoxication"] bacteriological origin." Massalongo tied the differing individual reactions to this underlying infection to provide evidence that each individual has a different predisposition for the site of infection. This bacteriological infection will result, wrote Massalongo, in either chronic or transitory tic symptoms depending upon individual predispositions.[73]

A Consensus on Psychopathology

Given the persistence of the connection between both Sydenham's chorea and convulsive tic behaviors with prior rheumatic disease, why did tic behaviors continue to be categorized as pathologically separate from other choreas? The answer to this question is tied, in large part, to the triumph of the psychologizing that increasingly affected late nineteenth-century thinking and developed into psychoanalytic categories by the 1920s. Ironically, the emphasis was on common underlying pathology rather than on variations in symptoms. For whenever it came to explaining the predisposing causes of convulsive tics, Sydenham's chorea, and hysteria, the Charcot circle, including its internal critics, were unified in their belief that these various tic disorders were the result of a set of vaguely defined inherited factors connected with theories of psychological degeneration. Their definition of psychopathology as hereditary degeneration rejected the possibility of specific localized neuropathology.[74] Instead, they suggested a more global, yet more vague, underdevelopment of the higher cerebral functions; one which left its victims with an arrested emotional development that both meshed with and reinforced infantile psychological behaviors.

This fixation on psychological theories of degeneration often led to ignoring clinical data that may have raised interesting alternative questions, especially in relation to the work of Koch and Pasteur, on the possible role of infection in these disorders. Thus, when his fifteen-year-old patient "G. D." volunteered that three months before his most recent recurrence of severe motor spasms and coprolalia he had contracted tonsillitis (a streptococcal infection), it never occurred to Gilles de la Tourette to ask G. D. if he had experienced similar illnesses prior to other outbreaks of convulsive tics.[75]

The assumption that motor and vocal tics were psychopathological and vaguely hereditary developed such paradigmatic force that practitioners felt obligated to explain away rather than to engage in analysis of any contrary evidence. For instance, Fulgence Raymond, who assumed Charcot's chair in 1895, used the tics and coprolalia of a thirteen-and-a-half year-old girl to show that her referring physician had made a diagnosis of Sydenham's chorea in "great ignorance."[76] Yet, that determination was made by her attending physician because prior to her onset of tics, the patient had been ill with "scarlatine, rheumatic fever, followed by an arthritic infection of her left knee," bacterial infections that many physicians then and today view as predisposing causes of Sydenham's chorea. Indeed, what Raymond might have noticed, as this case exemplified, was the often problematic classification of a number of childhood movement symptoms both in terms of underlying cause and the course of illness.

Raymond's failure to consider a possible bacteriological connection to movement disorders highlights the difficulty that practitioners face when confronted with evidence that, by its very nature, requires rethinking established classification boundaries. Even when these Salpêtrière physicians adopted a wider etiological view, they fell back on vague nineteenth-century neo-Lamarckian views of heredity. As a result, whatever their disagreements Salpêtrière physicians like Raymond continued to insist that the underlying cause of childhood vocal and motor tics was hereditary degeneration.[77]

Similar assumptions can be found in the first North American case of convulsive tics reported in 1886 in the *Journal of Nervous and Mental Disorders* by C. L. Dana and W. P. Wilkin. They described the case of a twelve-year-old male who, following a fall in which he sprained his foot, "would suddenly and involuntarily burst out into expressions of the most profane and obscene character; repeating them rapidly for a few moments and then stopping." Until that time, the child had never used obscene words, but now a sudden noise or anything that startled him would cause a rapid and automatic vocal outburst, most often of the word "shit!" Although the boy "did not seem to understand the significance of the words used," nevertheless, he was sent home from school. The boy also developed echolalia: "he would repeat not only the last words but the ends of sentences; and do it automatically imitating to a considerable degree the tone and accent as well as the words of the speaker." He had a compulsion to confess "things he wanted most to conceal." For instance, on one occasion he broke a dish and persuaded his brother not to tell their mother. Yet when the mother returned home, the patient "burst out: 'I broke the dish.'"[78]

Like Gilles de la Tourette and Charcot, Dana and Wilkin insisted that *tic convulsif* was distinct from chorea. Yet their report considers the anomalies between their patient's history and Gilles de la Tourette's typology. They noted that their patient's convulsive movements developed *after* his speech disturbances and that when these motor tics appeared, they almost entirely replaced the involuntary vocalizations and cursing. This, Dana and Wilkin pointed out, was the opposite of the order reported in Gilles de la Tourette's cases. They also claimed, again in contrast to Gilles de la Tourette's assertions, that medications, including bromides, arsenic, and iron, appeared to control the boy's symptoms. But Dana and Wilkin were so attached to widely-held degenerative assumptions, these contradictions failed to persuade them to examine the possible meanings of the anomalies they uncovered. Instead, they concluded that their patient's behaviors had no organic cause but rather constituted a "neuro-degenerative" disorder. "It is an evidence, like epilepsy and paranoia, of family decay."[79]

This type of thinking proved influential even as far away as Australia. In 1895 Australian physician R. R. Stawell, drawing on Charcot and Gilles de la Tourette, explained that habit spasm and convulsive tics shared "the etiology . . . that the affected children are as a rule neurotic" and "emotional . . . In some cases there is a distinct neurotic family history." Yet, Stawell explained, his patient was not neurotic, nor were any of his family. The only clinical issue of note was that "seven months prior to these attacks he had scarlatina," a streptococcal infection. Stawell, however, presented this as having no connection with the patient's tics.[80]

Similarly, in 1896 a Philadelphia pediatrician, J. C. Wilson, reported the case of a fifteen-year-old boy, Z., who "closely corresponds to several in the series published in 1885 by Gilles de la Tourette."[81] Wilson reported that the child, who was hospitalized, had mumps, measles, and, at age ten, an attack of malaria. At twelve Z. became "fidgety and restless," and soon after "developed spasmodic twitchings of muscles of his face, and had difficulty swallowing." The facial twitchings became more frequent and gradually affected the muscles of the neck, upper extremities, and thorax. General body paroxysms followed. Wilson also reported that Z.'s mind was affected: the boy "collected numerous small articles, playthings, spoons, and the like, which he took to bed with him." Z. also began making the involuntary "inarticulate sound—h-m, h-m-m, h-m-m-m." For the previous two years, Z.'s tics had been so intense that he was unable to attend school. Wilson included four photos of Z., illustrating the boy's movements from tics to paroxysm. Wilson opted for treatment with medications, especially sedatives, and placed Z. on a milk diet.[82]

When none of these medications seemed to work, a circumcision was performed because Wilson believed that the boy's tics were exacerbated by masturbation. The operation initially seemed to reduce the frequency and intensity of Z.'s tics, but the symptoms soon returned and so did Z.'s masturbatory habits. Deciding that [Z.] "was addicted to masturbation," Wilson ordered that the boy's arms and hands be put in splints at night. When Z.'s mother learned of these restraints she was furious and immediately removed Z. from the hospital and from Wilson's care. This confirmed Wilson's view that the mother was a "highly neurotic person with an uncontrollable temper." Wilson never saw the child again, but he believed that Z.'s case demonstrated Beard's and Gilles de la Tourette's claims that this was a progressive disease that moved through three stages, eventually manifesting itself in explosive cursing and "by an irresistible impulse to imitate the words or acts of bystanders." Because Z. had presented neither of these behaviors, Wilson concluded that Z. "was in the second . . . stage of violent muscular contractions with inarticulate sounds," but there was no doubt that Z. would develop coprolalia. "We recall the saying of Beard," concluded Wilson, that "once a jumper always a jumper."[83]

Because psychoanalysts later would connect motor and vocal tics to the *repression* of masturbatory urges, it is difficult not to notice this earlier association between *actually* masturbating and the eti-ology of tics.[84] Wilson's concern with Z.'s masturbatory habits reflects a generalized, especially Anglo-American, medical belief that masturbation resulted from and led to a series of degenerative mental disorders, sometimes referred to as "masturbatory insan-ity."[85] In 1901 Otto Lerch, Professor of Clinical Medicine at New Orleans Polyclinic Hospital, attached the severe tics and vocaliza-tions (but not coprolalia) of a fifty-five-year-old German immigrant to his youthful masturbatory habits, which weakened his constitu-tion, bringing on the tics. "The history of self-abuse practiced from childhood, the disgust for women, and the shy retiring disposition, which the patient . . . has always possessed, and the headaches occurring since he can remember without ascertainable cause" demonstrate, Lerch insisted, that "the disease in this case is a purely psychic condition characterized by the described involuntary spas-tic movements and is thought to be due to congenital malforma-tion."[86]

Lerch, who reported that the patient had measles and scarlet fever [a streptococcal infection], found it unremarkable that the patient's recurrence of involuntary tics and vocalizations initially appeared after "a severe cold during which he would sniffle and occasionally throw his head back; within a few months the habit was firmly established and has continued ever since, the spasms becoming more and more severe." The fact that bromides and other medications had no long-term impact on the disease served as fur-ther evidence for Lerch of the psychological etiology of the disor-der, whose "prognosis is always bad as to recovery."[87]

The anomalies seemingly so evident in Gilles de la Tourette's typology did not lead to a resurgence of theories linking tics and movement disorders with prior infections. In fact, by the early twentieth century the connection between convulsive tics, Syden-ham's chorea, and rheumatic fever almost totally disappeared. Thus in 1956 when Angelo Taranta and Gene H. Stollerman demon-strated the link between Sydenham's chorea and Group A strep, they were unaware that others in the 1890s had made similar claims.[88] The reasons for this historical amnesia are complex. Two related factors help explain the eclipse of the exploration of infec-tious connections with tic behaviors. First, those who subscribed to degenerative explanations for convulsive tics (whether they agreed with Gilles de la Tourette or Guinon) ignored any evidence connecting these behaviors with prior infections and specific local-ized neuropathology. Second, hereditary degenerative explanations for tic behaviors were validated with the publication of Henry Meige and E. Feindel's extremely influential 1902 *Les Tics et leur Traitement.*[89] This study, which was republished in English in 1907,[90] would profoundly influence all thinking about tic disorders for the next half century.

The Case of "O." and the Emergence of Psychoanalysis

··· 4 ···

By the early twentieth century, physicians who attempted to diagnose and treat patients presenting with tics and involuntary vocalizations were forced to draw from a confusing array of contradictory claims. Even subsequent issues of the same medical journal instructed practitioners that these behaviors fit Gilles de la Tourette's typology, or that they were a form of hysteria, or that tics were connected to rheumatic fever. Physicians desperately required a comprehensive and accessible clinical text that laid out competing claims and offered a guide for differential diagnoses. This need was met with the publication of Henry Meige and E. Feindel's 1902 study, *Les Tics et leur Traitement*, which was translated into English by the British neurologist S. A. Kinnier Wilson in 1907. *Tics and their Treatment* became the standard for diagnosis and treatment of motor and vocal tics for the next half century.[1] Meige and Feindel's explanation seemed to clear away the contradictions and anomalies first exposed by Georges Guinon and expanded by their mentor Edouard Brissaud. Building on the work of Brissaud, Meige and Feindel developed a comprehensive explanation for persistent tic and obsessive-compulsive behaviors that addressed both the variety of symptoms and the differing course and outcomes experienced by those displaying these symptoms. As important, they seemed to transcend the troublesome and confusing issue of whether motor and vocal tics were part of a continuum of hysterias or whether some tics were to be differentiated from

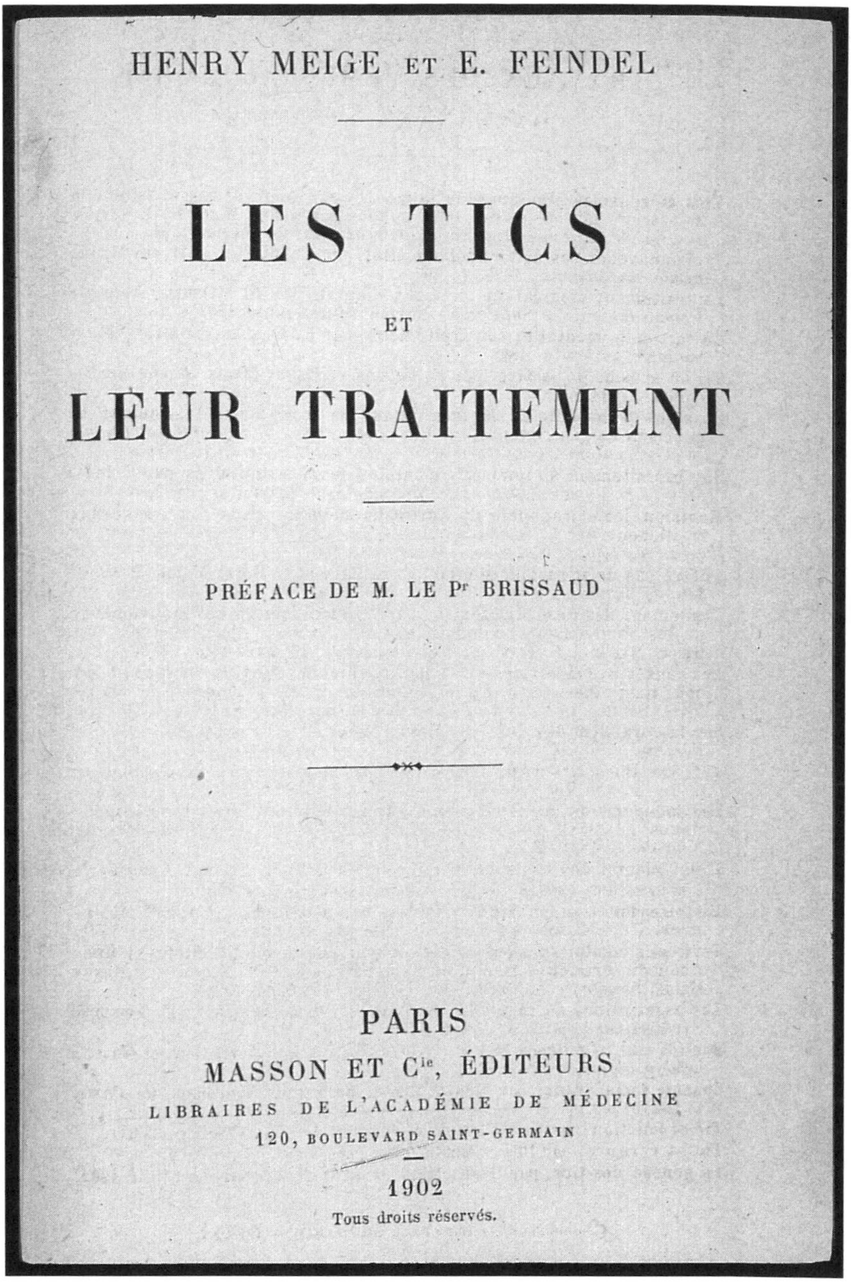

HENRY MEIGE et E. FEINDEL

LES TICS

ET

LEUR TRAITEMENT

PRÉFACE DE M. LE Pr BRISSAUD

PARIS

MASSON ET Cie, ÉDITEURS

LIBRAIRES DE L'ACADÉMIE DE MÉDECINE

120, BOULEVARD SAINT-GERMAIN

1902

Tous droits réservés.

Figure 4.1 The publication of Henry Meige and E. Feindel's 1902 study, *Les Tics et leur Traitement,* which was translated into English by the British neurologist S. A. Kinnier Wilson in 1907, became the standard for diagnosis and treatment of motor and vocal tics for the next half century.

TICS · AND
THEIR TREATMENT

By HENRY MEIGE and E. FEINDEL

With a Preface by
Professor Brissaud

TRANSLATED and EDITED, with a CRITICAL APPENDIX
By S. A. K. WILSON, M.A., M.B., B.Sc.
Resident Medical Officer, National Hospital for the Paralysed and Epileptic.
Queen Square, London

NEW YORK
WILLIAM WOOD AND COMPANY
1907

hysteria in terms of their alleged etiology and by the success or failure of interventions against them. Coming when it did, just as a newer psychological framework of explanation was beginning to make its appearance, *Les Tics et leur Traitement* proved to be a timely and convincing alternative to Charcot's and Gilles de la Tourette's typology.

Brissaud's introduction to *Les Tics et leur Traitement* elaborated a psychopathology of tic disorder. An examination of a ticcer "always reveals insufficiency of inhibition" due to the persistence of many "bad habits." Unless these bad habits were restrained, simple childhood tics would evolve to debilitating behaviors. The more a tic is repeated, claimed Brissaud, "the more inveterate it becomes, and the greater the likelihood of its becoming generalised." Repetition simultaneously could transform habits into persistent tics in those with an inherited predisposition to imitation. Thus, "the distressing neurosis described by Charcot and Gilles de la Tourette as the 'disease of tics,'" Brissaud insisted, "is no more than the superlative expression of a neuropathic and psychopathic disposition entirely akin to that favoring the development of the most harmless tic. Its earliest exhibition is a series of apparently insignificant bizarre convulsions; but its indefinite prolongation, its gradual involvement of one limb after another, its association with grave mental symptoms, and its frequent termination in dementia, are reason enough for eyeing the first little premonitory tic with mistrust, and combating it with vigor." Contrary to Gilles de la Tourette's assertions, Brissaud insisted that convulsive tics were curable if intervened against early and vigorously. The primary therapeutic aim, according Brissaud, was reinforcement of the patient's will.[2]

Reflecting their mentor's emphasis on hereditary etiology, Meige and Feindel dispensed with Gilles de la Tourette's typological boundaries. The latter had led to "confusion" due to their "unfortunate sacrifice of analytical accuracy to a premature desire for the schematic classification of disease."[3] Meige and Feindel's view proved compatible with the two strains of emerging explanations for a variety of seemingly psychopathological behaviors. On the one hand, it meshed with eugenics, which portrayed behaviors like motor and vocal tics as the result of degenerative inheritance. On the other hand, it segued into Freudian explanations of early childhood sexual repressive conflict. In 1921 Sandor Ferenczi, having no patient sample of his own, appropriated Meige and Feindel's cases as exemplars of what became the classic psychoanalytic explanation of tic behaviors.[4] Thus Meige and Feindel's claims, rather than Gilles de la Tourette's, framed the views and assumptions that European and American practitioners held about convulsive tics until the 1960s.

Les Tics et leur Traitement opens with "The Confessions of a Victim to Tic," an edited and annotated reproduction of a memoir written by a fifty-four-year-old successful businessman, referred to in the text only as O. He suffered a lifelong affliction of facial, body, and vocal tics combined with an assortment of compulsive behaviors, which Meige and Feindel considered to be prototypical of the tic.[5] The text and conclusions were actually completed before O. was discovered by the authors. Meige and Feindel had difficulty locating an archetype of the behaviors they had described, but in 1901 they located an individual "who is a perfect compendium of almost all the varieties of tic, and whose story, remarkable alike for its lucidity and educative value, forms the most natural prelude for our study."[6] Thus "The Confessions of a Victim to Tic" was edited and interpreted by Meige and Feindel to conform with and authenticate the authors' wider claims.

As clinicians' appropriation of the case of the Marquise de Dampierre demonstrated, such writing strategies are not unique in the history of the classification and treatment of tic behaviors. Emblematic cases are often recreated and interpreted *after,* rather than before, etiological theories have been generated. No doubt Meige and Feindel believed they were "objectively" reproducing a case history, but the process they employed, as well as the context they provided, resulted in only one among a number of possible readings of O.'s story. Like its predecessor case, "The Confessions of a Victim to Tic" would be appropriated and reconstructed by future theorists and writers as prototypes for their explanations of the causes for ticcing behaviors.

Meige and Feindel's reading of the case of O. is best understood in the context of their earlier research. It drew heavily on a series of articles that elaborated Brissaud's view that in a susceptible population tics resulted from habits formed during childhood.[7] Framing their view of tics was a larger set of beliefs that implicated hereditary degeneration as a central feature in a variety of behavioral psychopathologies. Meige's 1893 doctoral thesis, *The Study of Certain Neuropathological Travelers: The Wandering Jew at the Salpêtrière,* described a congenital illness that caused Jews to wander from place to place, making it impossible for them to form attachments to any place or nation.[8] Meige's analysis was based on his and Charcot's observations of Jewish patients at the Salpêtrière, supplemented by an assortment of popular writings about "Wandering Jews," including notoriously anti-Semitic texts. Their Jewish patients, wrote Meige, were "constantly obsessed by the need to travel, going from city to city, from clinic to clinic, in search of a new treatment, of a yet unknown remedy. They try every medication that anyone suggests, greedy for novelties; but they quickly reject them, inventing a frivolous pretext for no longer continuing them, and the impulse reappears, when one fine day they run off entranced by a new delusion of a distant cure." The characteristics displayed by Jewish patients were, according to Meige, merely a reflection of a deeper "quality of their race." Jews are "first of all profoundly neuropathological . . . And what's more they submit to irresistible impulses which trap them in a perpetual vagabondage. Their obsessive idea is not absurd to them; nothing is more legitimate than searching for lucrative work, or for an efficacious remedy."[9]

Similar observational methods informed Meige's claims of the congenital causes of ticcing behaviors. Persistent motor and vocal tics were evidence of a degenerative psychological disorder characterized by regression to infantile behavior. *All* tics, Meige insisted in a 1901 article, were psychological.[10] The patient's "mental state," he wrote in 1902, "played a central role in the genesis of the tic."[11] Even where the initial goal of tics was to react to or to avoid harm from some actual situation, the tic became habitual, after the cause and goal had disappeared, in those who had "a psychological predisposition that above all confirmed hereditary weakness of the will."[12]

Evidence for this hereditary "weakness of the will" was gleaned from an examination of a patient's familial pedigree, in which any of a number of conditions was portrayed as a hereditary cause of tics. O.'s family history, explained Meige and Feindel, confirmed "the existence of a grave neuropathic heredity, an unfailing feature of tic." His grandfather, who married a first cousin, was a stutterer and had facial and head tics; his brother also was a stutterer, while both his daughter and sister had facial tics, and a son was afflicted with asthma. Aside from his tics, O.'s general heath was excellent. He exercised regularly and he "maintain[ed] a vigour and agility above the average" and "his intellectual activity" was "keen."[13] Because, like O., many of their tic patients seemed to display extraordinary intelligence, Meige and Feindel adopted the view that these ticcers suffered from "superior degeneration," an inherited condition in which infantile regression was mixed with superior intellectual abilities.[14]

Meige and Feindel were critical of claims that specific localized cortical lesions were the cause of ticcing behaviors. "We must emphasise the fact once again," wrote Meige and Feindel, "that mental predisposition is a *sine qua non* for the development of a tic." Although they admitted it would be premature to rule out a strictly localized physiological pathology (lesions) for tics, Meige and Feindel concluded that the behavior they had elucidated "pertains to a psychical rather than to a motor sphere, and is to be regarded as a disease of the will."[15] They saw psychological degeneration as ultimately providing a way to connect the psychological and biological causes of tics: psychological underdevelopment (fixation at infantile behavior) was linked to the underdevelopment of cortical structures (hypoplasia). Tics resulted from "some congenital anomaly, some arrest or defect in the development of cortical association paths or subcortical" channels or "malformations that our medical knowledge is still unhappily powerless to appreciate."[16] Meige and Feindel explained that although most children eventually outgrew early ticcing behaviors, in a subset of children with a congenital predisposition like O., motor and vocal tics as well as obsessive behaviors persisted.

"Mimicry," wrote Meige and Feindel, "is strong in the child's nature, and bad habits are quickly contracted." If a child inherited "a nervous weakness" that child was especially susceptible to suggestion of all forms and therefore at great risk in developing tics.[17] O. was a mimic. "A curious gesture or bizarre attitude affected by anyone," O. reported, "was the immediate signal for an attempt on my part at its reproduction." O. remembered that when he was thirteen he saw "a man with a droll grimace of eyes and mouth, and from that moment I gave myself no respite until I could imitate it accurately." Similarly, O. attributed his head tossing movements to his attempt to mimic two school mates who habitually tossed their long hair back by a shake of the head. Although O. reported that he had never experienced an irresistible urge to curse, Meige and Feindel claimed that O.'s impulse to use slang was "a sort of *fruste* [substitute] coprolalia."[18]

Meige believed that the onset of tics could be connected to a response to an earlier pain or actual physiological event. The behavior persisted in those with congenitally weak self-control. O. reported that his eye and lip movements had their origins in reactions to actual stimulations, but they continued as if by habit long after initial stimulation ceased. He continued to sneeze and sniffle for months after falling snow had caused a tickling sensation. Such admissions, noted Meige and Feindel, were evidence of O.'s "pathogenic" lack of will. When O. reported that his head and face movements were due to the annoyance caused by seeing the tip of his nose or his moustache, Meige and Feindel argued that it was O.'s "force of repetition" that "changes the voluntary act into an automatic habit, the initial motive for which is soon lost; and the patient shows the weakness of his character by making little or no effort at inhibition." O. insisted that he could not control these desires: "There seem to be two persons in me . . . I am at once the actor and the spectator; and the worst of it is, the exuberance of the one is not to be thwarted by the just recriminations of the other."[19]

O. related his ongoing and futile attempts to suppress and hide his tics but lamented that these strategies failed because they too "become so habitual that they are nothing less than fresh tics appended to the old. To dissemble one tic we fashion another." In an attempt to restrain his head tics, O. resorted to a series of failed and increasingly desperate contrivances, from resting his head on a cane, to wearing stiffly starched collars, to literally tying his head to his trousers. Nevertheless, Meige and Feindel discounted O.'s rational strategies and instead reasserted their view that O.'s behaviors revealed that psychological weakness always preceded the motor movements. O.'s claim that he developed alternative strategies to suppress the unwanted tics was a ploy that hid the fact that the substitutes were merely similar sensations that took different forms. O.'s explanations underlined Meige and Feindel's belief in the infantile nature of the man's behaviors.[20]

Meige and Feindel found confirmation for their claims of a psychogenic pathology of O.'s tics in his ability to suppress his tics temporarily. "Should he find himself in the company of one from whom he would fain conceal his tics, he is able to suppress them

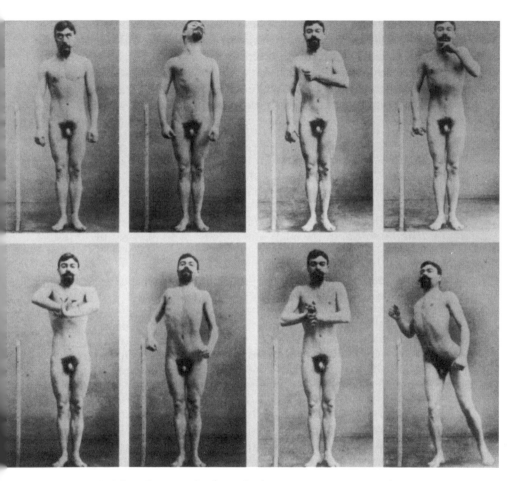

Figure 4.2 This photograph of a naked ticcing patient appeared in a 1906 article in a journal established by Charcot and edited by Henry Meige that used photographs of patients as diagnostic guides for physicians.

for an hour or two, and similarly if he is deep in an interesting or serious conversation." Yet after a while, "he can refrain no longer," and "he will invent any pretext for leaving the room, abandoning himself in his moment of solitude to a veritable debauch of absurd gesticulations, a wild muscular carnival, from which he returns comforted, to resume sedately the thread of the interrupted dialogue." As further evidence of O.'s ability to control his tics if he set his mind to it, Meige and Feindel pointed out that O. managed to cycle, and that "his devotion to billiards, or to such exercises as fencing or rowing, is never interfered with by an unruly tic. He is a great fisher, and . . . he will remain motionless indefinitely." If "interest in his prospective catch fade[d]," however, O.'s tics immediately returned.[21]

O. reported that his uncontrollable tics were like a "desire for forbidden fruit. It is when we are required to keep quiet that we are tempted to restlessness." When instructed to be silent in school, O. remarked that he sought "to evade the galling interdict by giving vent to some inarticulate sound. In this fashion did my 'cluck' come into being." Such statements confirmed Meige and Feindel's view that O.'s account of the origin of his tics supplied further evidence of a "mental infantilism," which prompted children to do exactly what they were forbidden to do. "They seem animated by a spirit of contrariness and of resistance; and if in normal individuals reason and reflection prevail with the approach of maturity, in these 'big babies' many traces of childhood persist, in spite of the march of years."[22]

Meige and Feindel found final confirmation for their characterization of O. in his reaction to recurrent abdominal pain, which was so acute that O. had to get up and walk about his room. Although Meige and Feindel did not doubt O.'s suffering, they were certain that it was attached to O.'s ticcing behaviors. They believed that such excessive reaction to pain was "characteristic of a nervous and badly trained child, not of a man of his years."

What is extraordinary about their judgments, as Meige and Feindel admitted, was that O. "has managed and still manages important commercial undertakings, demanding initiative and decision, and so far from sparing himself in any way, he has exhibited a combination of caution and audacity that has stood him in good stead." Instead of recognizing this as an adaptive strategy by which O. was able to channel his compulsive tics in the service of his vocation, Meige and Feindel cited O.'s successful business career as evidence that O. could selectively control his tic behaviors whenever he chose. O.'s failure to do so at other times, they insisted, proved that his tic behaviors were the result of willful bad habits and of a "mobile and impulsive temperament."[23]

Perhaps Meige and Feindel were so tied to their assumptions about the psychopathology of ticcers that they failed to appreciate the pain and hopelessness that O. attempted to convey. "In regard to my tics," O. wrote, "what I find most insupportable is the thought that I am making myself ridiculous and that everyone is laughing at me. I seem to notice in each person I pass in the street a curious look of scorn or pity that is either humiliating or irritating." Most of all O. wished to be inconspicuous and to hide his tics by any means available. But, he lamented, "nine times out of ten our efforts are abortive simply because we invent a tic to hide a tic and so add to the ridicule and the disease."[24]

Focusing on their preconception that heredity had combined with habit and weakness of will, Meige and Feindel denigrated the very source they drew upon. "Alike in speaking and in writing," wrote Meige and Feindel, "O. betrays an advanced degree of mental instability." Although they based their theoretical claims on O.'s written memoir and their interviews with him, Meige and Feindel characterized O.'s conversation as "a tissue of disconnected thoughts and uncompleted sentences." When O. reported that he had several times entertained thoughts of suicide, Meige and Feindel made light of the matter, informing their readers that "the suicidal tendencies of some sufferers from tic are seldom full-blown."[25] Encountering these similar comments, a reader might be puzzled by Oliver Sacks's statement that Meige and Feindel's discussion of the case of O. is "one of the finest examples of . . . collaboration I know." It reflects, writes Sacks, "a time in which patients and physicians still spoke the same language—and it was still possible for both to collaborate, producing between them a perfect balance of description and comment."[26]

O.'s treatment lasted only a few months and combined Brissaud's "'immobilisation of movements' with [Albert] Pitres's respiratory exercises and the mirror drill." Brissaud's method of "immobilisation of movements and movements of immobilisation" physically restrained O. from ticcing during sessions lasting up to half an hour and subsequently forced him to voluntarily exercise those muscles involved with the alleged locus of his tics.[27] By combining two forms of control, restraint and movement at will, Brissaud believed that ticcing patients could reverse their habit tics.[28] The second part of O.'s treatment was the breathing exercise developed by Charcot's former collaborator, Pitres (1848–1928).[29] Pitres's method rested on the supposed "physiological rapport which existed between movement and thought on the one part, and breathing on the other part."[30] O. was instructed to take and hold a deep breath for several minutes, twice a day. With his back and shoulders against the wall, he would raise and lower his arms eight times a minute, decreasing steadily to four a minute.[31] Finally, in Meige's so-called mirror drill, O. watched himself in front of a mirror for five minutes. Whenever he saw himself tic, he had to repeat the tic voluntarily several times. The purpose of this procedure, according to Meige, was to make the patient conscious of his tics and thus able to control them.[32]

"The patient," wrote Meige and Feindel, "has recovered his self-confidence, and the compliments of his friends prove an additional restorative." O.'s physicians admitted that "the tics still recur, but their number is less, their duration shorter, their severity considerably diminished." O. added that he "very much doubted whether I shall ever have the necessary perseverance to master all my tics, and I am too prone to imagine fresh ones; yet the thought no longer alarms me. Experience has shown the possibility of control, my tics have lost their terror."[33] Although the case of O. framed Meige and Feindel's text, their treatment of him lasted less than a year. They continued to write about tics and their treatments, but O. never reappeared in their articles. Thus, it remains impossible to learn to what extent O.'s reported improvement was permanent or merely another episode in a life filled with waxing and waning tics.

Meige and Feindel's claims were the focus of a special session of a meeting of psychiatrists and neurologists held at Grenoble in August 1902.[34] Participants included most of the important French tic researchers: Meige, Brissaud, Pitres, Joseph Grasset, and Jean-René Cruchet of Bordeaux. The discussion went back and forth with citations of patients' case histories as ammunition for conflicting claims. M. E. Noguès of Toulouse was skeptical about the connection of all tics with mental problems, especially labeling ticcers as infantile. Rather, Noguès believed that involuntary motor movements created both anxiety and confusion, and these made ticcers often seem as if they were mentally infantile.[35] Meige forcefully defended his view of the psychic infantilism of ticcers: "Mental trouble predominates" because the disorder is always connected with "weakness of volition" that, despite a ticcer's age, always manifests itself *"in a partial arrest of psychic development."*[36] Grasset of Montpellier disagreed with both Meige and Noguès. He was "repelled by the words *mental illness, mental alienation*" because psychic phenomena were complex and should be tied to the functioning of the brain.[37] Pitres disagreed with Meige and Feindel's classifications and definitions, and his student, Jean-René Cruchet, argued that Pitres's method "was superior to Brissaud's," especially "in the treatment of psycho-mental convulsive tic."[38] Cruchet, now chef du clinique at the Faculté de Médecine at Bordeaux, outlined his typology of tics as essentially clonic (involuntary activity of the nerves).[39]

In the following years Cruchet expanded his critique. In a 1906 article published in *Archives Générales de Médecine,* Cruchet challenged both the construction of habit tics and the putative mental conditions that Brissaud, Meige, and Feindel had attached to all convulsive tics.[40] A habit tic, Cruchet began, should not be confused with a convulsive tic because it was "comparable with a normal movement" and "only different from normal movements because *it was not wanted* at the moment when it was executed and that it was accomplished without reason or utility, with unusual frequency." In contrast, a convulsive tic was "an exaggeration of a habit tic, . . . the movements of convulsive tics are involuntary, absurd, urgent. The movement is involuntary because it is executed despite the volition of the subject."[41]

Cruchet's main aim, however, was to contest the claim that psychological degeneration underlay convulsive tics. Using the case of Raymond, a thirteen-year-old boy with a history of motor tics, Cruchet argued that what appeared to be mental infantilism was actually an irresistible impulse that the boy was unable to contain. Raymond's intelligence and judgment were sufficiently developed so that he feared punishment. Yet he lacked the ability "to stop him[self] from performing useless and absurd acts, of having pointless and foolish ideas, of pursuing disagreeable habits." Raymond also displayed what Cruchet labeled "mental feminism." "He has the manners and tastes of a woman, he detests noisy games and brutality, he is a little quarrelsome, likes to work in the kitchen, takes extreme pleasure in cradling babies in his arms, would like to be a 'soldier' because the uniform attracts him." Even though all of these behaviors could be read as evidence of mental illness, Cruchet pointed out that Raymond's convulsive tics waxed and waned without changing his mental state. This persuaded Cruchet "that convulsive tics constitute a particular muscular instability, which sometimes coexists with a defective mental condition, which singularly aggravates it, but which does not create it."[42] Thus, Cruchet rejected assertions that convulsive tic symptoms were evidence of insanity.

When the English-language version, *Tics and their Treatment*, appeared in 1907, the translator, British neurologist S. A. Kinnier Wilson, added an appendix to respond to Cruchet's criticisms. Wilson rejected Cruchet's assertion that tics resulted from nerve disease and that the differences between Cruchet and Meige and Feindel were about labels rather than processes.[43] What Cruchet called "habit tic" was equivalent to what Meige and Feindel described as "stereotyped acts." Cruchet's denial of a connection between the mental state of a person and his or her ticcing behaviors was, in Wilson's view, simply a matter of semantic difference.[44]

Wilson's translation of *Les Tics et leur Traitement* made Meige and Feindel's views available to a wide cross-section of the Anglo-American medical community, while Cruchet's work was available in English only through the filter of Wilson's appendix. Anglo-North American and continental clinicians drew on as well as adapted Meige and Feindel to construct a physio-psychological set of explanations that connected bad habits with motor and vocal tics and obsessive behaviors.[45] Most often, as with Meige and Feindel, these symptoms were also tied to nonspecific hereditary factors.[46] Some physicians, particularly in North America, relied on the all-purpose diagnosis of "neurasthenia" (nervous exhaustion) as a vehicle for incorporating Meige and Feindel's explanations with emerging psychoanalytic claims about the origins of tics.[47] These discussions of tic had often quoted Freud, suggesting that psychoanalysis could be made compatible with the claims of Meige and Feindel.[48]

Enter Psychoanalysis

In 1921, drawing on the case of O., the Hungarian psychoanalyst Sandor Ferenczi (1873–1933) outlined the first purely psychoana-

Figure 4.3 In 1921 the Hungarian psychoanalyst Sandor Ferenczi, drawing on Meige and Feindel, outlined the first purely psychoanalytic analysis of tics. Ferenczi concluded that tics were a symbolic expression of masturbation. This became the official psychoanalytic view of tics.

tics resulted from repressed masturbatory desires.

lytic analysis of tics.[49] Ferenczi's paper, "Psycho-Analytical Observation on Tic," became the official psychoanalytic statement on tics.[50] Yet, Ferenczi reached his conclusions without ever examining a single ticcing patient. Instead, he relied entirely on the reports in Meige and Feindel's 1902 *Les Tics et leur Traitement,* particularly on their description, reproduction, and discussion of O.'s "Confessions of a Ticqueur." Ferenczi sought to justify his omission of actual clinical cases by claiming that he had few opportunities to personally observe ticcing patients in his private practice. Relying on Meige and Feindel, he could not be accused of biased observations or of making suggestions to a patient.[51]

Although all medical researchers tend to shore up their claims by referencing the work of important predecessors, with psychoanalysis this practice required turning to Freud for validation. Ironically, Freud himself had said little explicitly about tics. His most extensive examination of convulsive tics, the case of Frau Emmy von N., which appeared in his and Josef Breuer's *Studies in Hysteria* (1893–1895), took place *before* Freud developed his theory of psychoanalysis.[52] Freud treated Frau Emmy von N. in 1888 or 1889.[53] At this time he accepted Guinon's view that motor and vocal tics were common symptoms of traumatic hysteria.[54] As previously pointed out, by the late 1880s both Charcot and Gilles de la Tourette had conceded Guinon's point that hysterics often displayed motor and vocal tics along with coprolalia. Whether a patient should be diagnosed as hysterical or afflicted with convulsive tics depended on whether or not the symptoms could be controlled by interventions such as hypnosis.

It was within this context that Freud treated Frau Emmy von N. In his description of the forty-year-old woman's symptoms, Freud reported that, along with frightening hallucinations, the patient stammered, displayed "frequent convulsive *tic*-like movements of her face and the muscles of her neck," and "frequently interrupted her remarks by producing a curious 'clacking' sound." Added to these, every two or three minutes the woman suddenly "broke off" her conversation, "contorted her face into an expression of horror and disgust, stretched out her hand . . . spreading and crooking her fingers and exclaimed, in a changed voice, charged with anxiety: 'Keep still!—Don't say anything!—Don't touch me!'"[55] Freud's diagnosis was hysteria resulting from repressed early childhood trauma that resurfaced due to her husband's death and subsequent legal conflicts with her husband's relatives. Freud used hypnosis as a vehicle for Frau Emmy to recollect her repressed early traumas. He encouraged Frau Emmy to reexperience a repressed traumatic event while she was in a hypnotic state and then instructed her to disregard it before waking. After eight weeks of treatment Freud claimed that the woman had improved rapidly.

Within a year, a stammering, clicking Frau Emmy returned to Freud's care. "A fresh psychical shock"—surprising a man in her maid's bedroom—had triggered the relapse.[56] Freud explained that Frau Emmy "had got into the habit of stammering and clacking whenever she was frightened, so that in the end these symptoms had come to be attached not solely to the initial traumas but to a long chain of memories associated with them, which I had omitted to wipe out."[57] After a second round of treatment, however, Freud was satisfied that "the therapeutic effect of these discoveries under hypnosis was immediate and lasting."[58] A year later he visited Frau Emmy von N. at her estate and reported that "she had only been disturbed by slight neck-cramps and other ailments" which Freud viewed as organic disorders. But, Freud admitted that while his patient was "in very many respects healthier and more capable, . . . there had been little change in her fundamental character."

Frau Emmy had developed a new phobia to riding trains, which Freud interpreted as a way "to prevent her making a fresh journey to Vienna" and his consulting room. The two never met again.[59]

Freud was familiar with Gilles de la Tourette's 1885 article and with Guinon's criticisms. In an essay, "A Case of Successful Treatment by Hypnotism" (1892–1893), Freud explained the theoretical difference between a diagnosis of hysteria and that of convulsive tics, but noted that in practice the distinction was generally arbitrary: "*Tic convulsif* [is] a neurosis which has so much symptomatic similarity with hysteria that its whole picture may occur as a part-manifestation of hysteria. So it is that Charcot, if I have not completely misunderstood his teachings on the subject, after keeping the two separate for some time, could only find one distinguishing mark between them—that hysterical *tic* disappears sooner or later, while genuine *tic* persists."[60]

Like most observers in the 1890s, Freud rejected Gilles de la Tourette's formulation of maladie des tics as a separate disorder and adopted Guinon's view that motor and vocal tics—that is, convulsive tics—were part of the spectrum of hysterical symptoms. Because Freud was convinced that his hypnotic cathartic treatments had ameliorated Frau Emmy's symptoms, he opted for a diagnosis of hysteria rather than "tic convulsif." But, even if Freud had concluded that Frau Emmy's symptoms were persistent, he nevertheless would not have felt obligated, any more than Guinon and others had, to resurrect the disputed designation "la maladie des tics de Gilles de la Tourette."

Freud emphasized Guinon's connection between compulsive ideation, such as the need to count to ten, and ticcing behaviors. "The picture of a severe *tic convulsif* is, as we know," wrote Freud, "made up of involuntary movements frequently (according to Charcot and Guinon, always) in the nature of grimaces or of performances which have at one time had a meaning—of coprolalia, of echolalia and of obsessive ideas." Guinon had found that his tic "patients reported that on some particular occasion they had seen a similar tic, or a comedian intentionally making a similar grimace, and felt afraid they might be obligated to imitate the ugly movements. Thenceforward they had actually begun imitating them."[61] Charcot had repeated, until his death in 1893, that hysterics had developed their tics through imitation or earlier trauma, while those with maladie des tics had not. Imitation and earlier trauma, in fact, were what Freud claimed he had uncovered in the case of Frau Emmy. Freud was, as he insisted, following the teachings of Charcot and Guinon.

The case of Frau Emmy was completed before Freud developed his psychoanalytic psychiatry. After 1893, however, Freud never reexamined the possible psychoanalytic mechanisms that might underlay tics and involuntary cursing.[62] Thus, Ferenczi and later psychoanalysts were forced to interpret the few contradictory hints that Freud had offered in passing conversation in a way that was congruent with psychoanalytic categories. "When I incidentally discussed the meaning and significance of Tic with Prof. Freud," wrote Ferenczi in 1921, "he mentioned that apparently there was an organic factor in the question."[63] Ferenczi reinterpreted Freud's remark to mean that tics were psychological: Freud, explained Ferenczi, should be understood as suggesting that particular organs serve as a "psychical representative" of a repressed conflict rather than as a physiological site for tic production. Tics were psychogenic conflicts that manifested themselves physiologically, as body movements and vocalizations.[64] Having demonstrated his allegiance to the master's insight, Ferenczi drew on Meige and Feindel's arguments and examples, translating them into a psychoanalytic vocabulary.

On the basis of Meige and Feindel's descriptions, Ferenczi concluded that tics were "stereotyped equivalents of Onanism [masturbation]" and that the connection of tics with eruptive cursing (coprolalia)[65] was "nothing else than the uttered expression of the same erotic emotion usually abreacted [released by acting out] in symbolic movements."[66] A number of earlier observers, including J. C. Wilson (1897) and Otto Lerch (1901), had connected masturbation with tics and coprolalia, *but* they had argued that ticcers were *actual* masturbators.[67] Turning this notion on its head, Ferenczi and his followers argued that tics resulted from *repressed* masturbatory desires. According to Ferenczi tics were best understood as a form of *"constitutional narcissism,"* where *"the smallest injury to a part of the body strikes the whole ego."* Ticcers were overly sensitive and their tics were the "motor expressions" of this hypersensitivity to external stimulation. The "hyperaesthesia" [hyperkinetic movements] were an "expression of narcissism, the strong attachment of the subject to himself, his body, or a part of his body," which Ferenczi tied to Freud's notion of "the damming-up of organ libido [energy associated with instinctual drives]." That is, these patients had repressed their wishes for sensual bodily pleasures, especially, but not only, those attached to their genitalia. Coprolalic outbursts and the inability to restrain a thought were, for Ferenczi, a substitute release of energy in those incapable of enduring a stimulus to their body without an immediate defense reaction. Involuntary cursing and blurting out other inappropriate speech were motor reactions through which these repressed emotions toward organ stimulation were released.[68]

Ferenczi was aware that tics often appeared after an infection, but this, too, he connected back to an injured libido. A physical illness, such as an eye infection, in patients with dammed-up libido led to a subsequent eye tic because of the unconscious defense against the infection's stimulation of the eyelids. The resultant tic was the psychic reaction to the physical stimulation caused by the infection: The tic symptoms displayed by these patients could be "traced back to the 'traumatic' displacement of libido and, . . . the motor expression of Tic arises from defence reactions against the stimulation of such parts of the body."[69]

If, as Ferenczi insisted, all tics had the same root, a reaction against a repressed libido, how were tics to be distinguished from their seeming opposite manifestation of catatonia, a rigidity and lack of movement, which had been explained by Freud as also resulting from dammed-up libido? The answer, in classic psychoanalytic style, was that these were two opposite outcomes resulting from the same underlying causes. Tics and catatonia, wrote Ferenczi, "had a common constitutional basis" that "explains the broad similarity of their symptoms." There was "an analogy between the principal symptoms of Catatonia—negativism and rigidity—with the immediate defence against all external stimuli by means of convulsive movement in Tic." Thus, Ferenczi asserted that the same underlying mechanisms that led to maladie de Gilles de la Tourette could conversely result in catatonia because both conditions are psychological reactions to repressed organ stimulation.[70] Although "the tic malady attacks children as a rule in the sexual latency period," this initial attack "can have various outcomes, apart from remissions, remaining stationary or degenerating into the symptom-complex described by Gilles de la Tourette." Alternatively, this motor oversensitiveness could be compensated for later in life by external inhibition of motor movements, that is as catatonia.[71] The imitative behaviors exhibited by tic patients provided, in Ferenczi's theory, another dialectical connection between tic and catatonia: "In schizophrenic Catatonia absolute mutism alternates with uncontrollable compulsion to talk and with Echolalia."[72] For Ferenczi then, the underlying pathology of those displaying motor and vocal tics was no different from the pathology of those who could not move or speak at all.[73]

Whatever else it did, Ferenczi's explanation transformed Gilles de la Tourette's classification into a psychoanalytic category in which motor tics and involuntary vocalizations were only one set of possible outcomes of a narcissistic, repressed childhood sexuality. Ferenczi found the evidence he needed in Meige and Feindel's "Confessions of a Tiqueur," the case of "O." Quoting O.'s statement that "I must admit that I am full of self-love and am particularly sensitive to blame or praise," Ferenczi found that this confession "show[s] tic patients as of a mentally infantile character, narcissistically fixated, from which the healthy developed part of the personality can with difficulty free itself." The fact that the vocal outbursts of these patients often had a sexual or "organ-erotic (perverse)" content was further confirmation for Ferenczi of their narcissism. He endorsed Meige and Feindel's conclusion that "every tic patient has the mind of a child." They are "big, badly brought-up children accustomed to give way to their moods never having learned to discipline their wills."[74]

However, Ferenczi rejected Meige and Feindel's explanation that these behaviors resulted from a "predisposition" or from "degeneration." "Where the patient can offer no explanation for the tic, they [Meige and Feindel] regard it as 'senseless and without purpose.' They forsake the psychological path too soon and lose themselves in physiological explanations."[75] In particular, Ferenczi claimed, Meige and Feindel neglected the unconscious sexual component that underlay all tics. Tics, for Ferenczi, increased in power during early puberty, pregnancy, and childhood, at the time "of increased stimulation of the genital regions." This explained why vocal outbursts often developed "into anal-erotic obscenities." For Ferenczi, these behaviors were a "displacement from below upwards"; with repressed sexuality, especially repressed masturbatory urges, manifesting themselves as a tic. The "excitability" of ticcers and their "tendency to rhythmical rubbing" was further evidence of the "genitalisation" of tics, which in some cases included "definite orgasm."[76]

Ferenczi's explanations formed the bedrock of all future psychoanalytic claims about the causes of tics and involuntary cursing. Like Ferenczi's attempt to find legitimation in Freud, others returned to Ferenczi's 1921 statements with a reverence generally reserved for the master himself.[77] Yet, ironically, Ferenczi's claims were based on the cases and writings of Charcot's descendants, Meige and Feindel, rather than Freud. Like Gilles de la Tourette, Ferenczi erected his typology and conclusions on a patient he had never examined. Following Ferenczi, a generation of psychoanalysts would frame their diagnosis and treatment of actual patients on an emblematic patient whose psychoanalytic diagnosis was based on textual interpretation rather than on clinical interaction. This may help us understand why, in the decades that followed, even many psychoanalytic psychiatrists opted for more eclectic etiological scenarios when their patients' symptoms and case histories could not be made to fit the story told by Ferenczi

Competing Claims

5

Meige and Feindel's book, supplemented by Ferenczi's views, influenced the diagnosis and treatment of tics well into the 1940s. But psychological claims continued to be challenged by evidence implicating physiological causes of tics. A worldwide epidemic of infectious encephalitis (1918–1926), in which tics were a common sequel, supplied persuasive support for an organic cause of these behaviors. Additional evidence came from reports that surgical removal of infected organs, such as sinuses or tonsils, relieved a patient's tics.

One solution to the contradictory claims of psychoanalysis and postinfectious theories was to separate patients for treatment based on examination of their case histories. Those who revealed a correlation between infection and subsequent tic could be treated biologically, while those without organic disease could be candidates for psychoanalysis. However, most patients' case histories were replete with evidence that could be marshaled to support whatever assumptions an examining physician brought to the clinical encounter. As a result, patients with similar medical histories and symptoms received different treatments depending on the philosophy and training of their particular physicians.

No matter what form of intervention they employed, physicians often reported success in ameliorating and even curing their patients' symptoms. In retrospect many of these claims appear to have resulted from a combination of wishful thinking, short follow-ups, and the tendency of tics to wax and wane. Given that tics can arise from a number of different underlying factors, however, some of the interventions may have been efficacious. In addition, the definition of what constituted tic disease was far from unanimous. Therefore, what was considered a cure depended on what symptoms and course were seen as essential elements of the disorder.

In 1922 psychiatrist Charles Trepsat published an article in which he claimed to have successfully treated a twenty-seven-year-old man through psychoanalysis. The man, Paul C., manager of an engineering company, had suffered from intense motor tics for twelve years, since the age of fifteen. Trepsat described his first encounter with Paul C.: "In the middle of a conversation, he suddenly stops speaking, his upper and lower right limbs forcefully contract and extend. The right half of his body moves in some sort of rotation before moving to the left. At the same time his head turns left and in concert with his eye balls which move in the same direction."[1] When Paul was absorbed in his work, he could sometimes go for a half an hour without manifesting tics. Also, he reported that he was able to conceal his tics from workers, but not from the company's director or from his own wife.

Although Trepsat did not cite Ferenczi's article of the previous year, he reached a number of similar conclusions in his diagnosis of Paul C.'s condition. In the psychoanalytic tradition, Trepsat searched for clues for Paul's affliction in his familial relations. The mother, according to Trepsat, was "neurotic, irritable, impressionable" and Paul's relationship with her was troubled.[2] The father died of cancer when Paul was sixteen. The year preceding his father's death, Paul had developed a secret romantic relationship with a local girl, whose parents, when they discovered it, forbade Paul and Anna from seeing one another. Nevertheless, for the next eight years the two secretly corresponded and exchanged poetry. During this time Paul developed spasms in his right hand. Little by little these spasms generalized to his arms, shoulders, legs, and trunk. The doctors at his lycée diagnosed Paul as hysterical and, at age eighteen, he was sent to Paris to consult with Charcot's former student Joseph Babinski; nothing positive resulted.

Under Trepsat's regimen of free association and dream interpretation, Paul revealed a deep fear of punishment for repressed wishes and imagined sexual transgressions. These included his early child-hood fantasy of marrying his mother; an incident in which, at age six, Paul and a girl his age had exposed themselves to one another; Paul's frequent masturbation; and his secret relationship with Anna.

Trepsat's psychoanalytic interpretation followed classic lines: Paul's fear of punishment for his fantasies and transgressions was manifested in the form of tics. Each tic and exacerbation served the dual function of revealing a repressed issue while simultane-ously punishing Paul for forbidden wishes or behaviors. Paul's initial hand tremors revealed both his fear that his secret corre-spondence with Anna would be discovered and a symbolic repre-sentation of the correspondence. Other tics reflected Paul's guilt over his masturbatory habits, while his dreams revealed Paul's deeper oedipal fantasies. These dreams served, in an overdeter-mined fashion, to conflate his obsession for Anna with his fantasies about his mother. In this context, Paul believed that his father's death was punishment for his (Paul's) sexual "crimes." Trepsat re-ported that as word association and dream interpretation brought each of these issues to Paul's consciousness, he was able to stop the tic associated with each repressed wish. What Paul's case proved, wrote Trepsat, was that "the origin of a tic can always be found in psychic repression."[3]

Contrary to Trepsat, other clinicians, including some of the most ardent supporters of psychoanalysis, proved extremely reluctant to attribute the underlying pathology of motor and vocal tics to psy-choanalytic categories. Most prominent among these clinicians was Swiss psychiatrist Raymond de Saussure (1894–1971), probably the most fervent advocate of Freudian psychoanalysis who was writing in French from the 1920s until his death in 1971. Son of the influential linguist Ferdinand de Saussure, Raymond de Saus-sure traveled to Vienna in 1920 to attend Freud's lectures and to be analyzed by him. In 1922, Saussure, won over by psychoanaly-sis, published his doctoral thesis *La Méthode Psychanalytique,* a volume whose aim was, as Freud wrote in its preface, "to give French readers a correct idea of what psychoanalysis is and what it contains."[4] In 1927 Saussure helped found the Paris Psychoana-lytic Society as well as the *Revue française de Psychanalyse.* Sub-sequently, Saussure served as vice-president of the International Psychoanalytical Association and president of the European Psy-choanalytic Federation.[5]

Yet, in 1923, when faced with explaining and treating convulsive tics, Saussure rejected Ferenczi's analysis in favor of an eclectic approach that combined organic (infectious) factors with the role of inherited infantile constitutions. Saussure was particularly influenced by the accumulated evidence that tics were one of the common sequels to infectious encephalitis.[6] At the midpoint of the encephalitis pandemic, 1923, one person out of every 1,000 in Europe and North America was infected and more than 20 million had died.[7] Often those who survived reported strange uncontrolled movements including tics.

Saussure's conclusions were formed by his treatment of a twenty-one-year-old ticcing patient named Blanche.[8] Born in July 1899 to an "extremely neurotic" mother and "an unknown father," Blanche, at eighteen months, was adopted by a farm family. The adoptive family included a grandmother, mother, father, and three girls who were all younger than Blanche. Although the grandmother was often overbearing, Blanche's adoptive parents were very affectionate and her stepsisters accepted Blanche as if she were a natural member of the family.[9] Blanche was an academically good and well-behaved student; at home she was affectionate and industrious. When Blanche was twenty, she was hired as a servant in the home of Madame M. in a nearby town.

In February 1920 Blanche suffered from mild flu, which left her extremely fatigued. She began to have nightmares in which she saw an old dead woman wearing a nightcap, who, despite the fact that her eyes were closed, stared at Blanche through the window. After these nightmares Blanche became extremely frightened and began to display "psychomotor agitations," consisting of convulsion of her limbs and trunk. She hallucinated that she was surrounded by men's heads, which were detached from their bodies. Blanche reported that people were following her, and she also babbled throughout the night. Madame M. called a doctor who sent Blanche home to her adoptive parents.

Once home, Blanche's hallucinations disappeared, but her motor movements persisted. She was generally agitated and could not sleep at night. During the day, however, Blanche could not stay awake and would fall into "a leaden sleep." After three weeks, she seemed to improve and, although her insomnia persisted, Blanche returned to Madame M.'s, where she worked without interruption until May 1921. "The insomnia led to an alteration in her personality which manifested itself above all by irritability and by a diminution of her memory."[10] During this entire time she also experienced menstrual cessation (amenorrhea). In early May, upon learning that Madame M.'s son had injured his leg in a motorcycle accident, Blanche began "to tremble like a leaf" and continued to do so for an entire day. Several days later she began to yawn about every three minutes, sometimes even more frequently. This new symptom was accompanied by an intense desire to sleep, especially during the day. The yawns were preceded by a feeling of constriction of her throat and chest, which produced a very strong pain in the joints of her jaw. Blanche soon developed other tics including increasingly frequent facial, neck, head, and mouth spasms. During some of her spasms, Blanche would cease breathing for a few seconds and then would breathe rapidly, as if in an attempt to counteract her involuntary yawning. Atrocious pain accompanied these spasms. The pain seemed to diminish by August, but she developed coprolalia and other involuntary vocalizations. Hypnotism was attempted, but it had no effect on Blanche's coprolalia or motor tics. On December 17, 1920, Blanche's parents sent her to the asylum at Cery-Lausanne where she was placed under Saussure's care.

Saussure first turned to psychoanalysis, but quickly ruled out hysteria. In fact, beyond her tics, Saussure was most struck by Blanche's difficulties with memory, especially because prior to her illness, Blanche had been reported to have demonstrated exceptional memory skills. Additionally, Saussure found that Blanche's symptoms seemed exacerbated by stress. For instance, when she was allowed a brief leave from the asylum to visit her adoptive family, Blanche became extremely agitated, especially after she learned that her savings had been used to pay for her stay in the asylum. After Blanche returned to the asylum, her tics had diminished, but her personality seemed permanently altered. She frequently got into arguments with other patients, became more anxious, believed she had been abandoned by everyone, and cried at the slightest provocation. Any stress increased her involuntary movements. Finally, Blanche's menstrual periods remained irregular and she reported gastric pains after eating. Blanche was released from the asylum in July 1922 and never was heard from again. If Saussure subsequently examined Blanche, he failed to report the results.

Saussure was convinced that Blanche's behaviors "did not constitute the symptomatology of a known psychosis," but "to the contrary, this symptomatology recalled all at once the alterations of character which had been described following epidemic encephalitis."[11] Saussure was reluctant to attach all of Blanche's symptoms to encephalitis. Rather, he believed that Blanche's illness should be viewed in the context of cofactors set in motion by the encephalitis. He concluded that the encephalitis had provided the soil for Blanche's congenitally weak will, giving rise to her ticcing symptoms. The encephalitis accounted for fatigue and pain. Blanche's initial yawning was a reaction to both of these, especially the lower jaw pain to which yawning was a direct reaction. But after the pain abated, Blanche's yawning continued, now as a habit, as "a tic in the sense of Meige." Other tics followed because Blanche was, in Saussure's words, a "dégénérée as described by Meige and Feindel . . . Her mother was extremely nervous, her unknown father was probably defective." Saussure was convinced that Blanche's reactions were infantile. For instance, she cried for an entire day when her bed was moved from the observation area to a different part of the hospital. Blanche found reading boring. "She is emotive," reported Saussure, and "has little will power."[12]

Neither Saussure's loyalty to psychoanalysis nor his familiarity with both Freud's and Ferenczi's views persuaded him that a psychoanalytic explanation of Blanche's behaviors was appropriate. Perhaps this explains why in his introduction to Saussure's *La Méthode Psychanalytique,* Freud distanced himself from Saussure's interpretations: "Since Dr. de Saussure has said in his preface that I have corrected his work, I must add a qualification; my influence has only made itself felt in a few corrections and comments and I have in no way sought to encroach upon the author's independence. In the first, the theoretical part of the work, I should have expounded a number of things differently from him . . . Above all, I should have treated the Oedipus complex far more exhaustively."[13]

German Somaticism

Whatever issues and claims separated Saussure's somaticism from Ferenczi's psychoanalysis, both approaches relied on and acknowledged their debt to Meige and Feindel. In contrast, German somatic psychiatrists, who did not have national, linguistic, or intellectual roots in the Salpêtrière tradition, drew on the more biological reductionist insights of Alois Alzheimer and Emil Kraepelin. Unlike Saussure, these practitioners saw no need to temper their organic explanations of the etiology of tics. Most influential of these was psychiatrist Erwin Straus at Berlin's Charity Neurological Clinic, who was convinced that "the organic nature of . . . compulsive tic diseases" can "no longer be disputed."[14] In two articles published in 1927, Straus, like others before him, remarked that motor tics and coprolalia were often a sequel to the "disease called chorea," which he believed to be caused by an "infection."[15] He based his conclusions on examination of fifty-eight ticcing patients (thirty-two males and twenty-six females) whose symptoms appeared between the ages of ten and fifteen years and whose case histories he laid out in great detail. The majority of them developed tics as a sequel to encephalitis. These infections, according to Straus, damaged portions of the brain's basal ganglia, resulting in the hyperkinetic movements and involuntary vocalizations of Tourette's disease. Straus found support for his analysis in a 1925 microscopic examination that revealed alterations in the basal ganglia of the brain of a patient whose tics followed an encephalitic infection.[16]

Building on these findings, Straus explained how coprolalia was connected to involuntary motor disturbances. His example was a twenty-eight-year-old male with motor tics, coprolalia, echolalia, and compulsive spitting, whose onset of symptoms began at age six after the boy was diagnosed with a mild case of chorea. After eight month's hospitalization the patient's motor disturbance seemed to have been cured. However, three years later, new uncontrolled motor movements occurred, which initially were diagnosed as compulsions. Soon after this the boy began to display uncontrollable movements and explosions of inarticulate sounds. A variety of treatments were attempted, but none proved effective. The inarticulate sounds and screams were gradually replaced by the compulsive utterance of obscene words, which was accompanied by "disturbances of respiration, phonetics and articulation." The symptoms fluctuated with short and long remittances. Although his fundamental personality was unaffected, the boy suffered considerably because his symptoms made socializing with others impossible. During clinical observation, the patient became easily irritated and excited and displayed suicidal tendencies. At times, the patient's "irritation increased temporarily into an intense fury and an urge to self-destruct," which required great effort by others to restrain. His obscene outbursts became more frequent and seemed to change situationally. Initially, when excited, the patient would blurt out ordinary words and expressions. Over time, however, all the outbursts were curses and usually were accompanied by involuntary motor movements. In addition, the patient frequently uttered inarticulate sounds which often evolved into screams.[17]

Straus wrote that it made neurological sense that the patient developed coprolalia because involuntary cursing resulted from the same underdevelopment that caused his motor disturbances. Thoughts and internal conversations are acquired in early childhood and both are tied to the normal development of motor control. Retardation of motor control affects the ability to restrain thoughts, which, as a result, are blurted out rather than censored. This tendency is especially exacerbated when a person becomes emotionally involved in a topic. Similar breakdowns occurred in everyday life when people became angry, or fought, or found themselves in danger. In these cases normal expressions increasingly were transformed into interjections. These might take the form of unarticulated sounds and, under greater stress, of swearing. The articulation of a word, Straus explained, merely served to complete the action of the unrestrained emotion. For Straus, not only was coprolalia analogous to motor tics, but also it actually resulted from the breakdown of the same brain system of motor movement restraints (the basal ganglia) that Straus had connected in his fifty-six other cases with a prior infection.[18]

Toward a Synthesis

The same year that Straus's two articles appeared (1927) the British neurologist S. A. Kinnier Wilson,[19] who two decades earlier had translated Meige and Feindel's *Les Tics and leur Traitement* into English,[20] organized a special session at the British Medical Association Annual Meeting on "The Tics and Allied Conditions." The session was attended by some of the most important clinical observers of tic from Britain, France, and the United States and underscored the resilience of psychoanalytic psychiatry, despite the challenges posed by the evidence of the role of epidemic encephalitis in the production of tics. Wilson's talk, along with commentators' reactions, appeared later that year in *The Journal of Neurology and Psychopathology.*[21] Wilson acknowledged that a number of observers had attempted to connect the origin of tics to encephalitis and other prior infections, but, as he warned his audience, such claims were "beset with difficulties, and, while it is natural to attribute them to the disease, we must not lose sight of the psychopathic soil in which they may be sprouting, for they are neither universal nor invariable as a sequella."[22]

Like Meige and Feindel and Ferenczi, Wilson's emblematic example of tic was the life history of O., with his constant exchanging of new tics for old ones. O.'s history, explained Wilson, demonstrated that "no prognostic significance attaches to the simplicity, complexity, or multiplicity of the tics in a given case; it is the patient's volition which counts."[23] In all cases of tic Wilson found "a constant component of a wretched family history of mental instability and degeneration . . . In no clinical class are tics more rampant than in idiots and mental defectives of all sorts."[24]

Wilson located the etiology of tics in psychoanalytic constructions. "Behind all tic phenomena," claimed Wilson, "lies a *psychical predisposition.*" Meige and Feindel's insistence in tying the etiology of tics to a mental process was, according to Wilson, substantiated by later observers, particularly by Ferenczi, who had argued that the ticcer was mentally infantile, or in psychoanalytic language, "narcissistically fixed." Wilson reported that his own experience confirmed that of Meige and Cruchet, which viewed the temperament of the ticcer as similar to that of the hysteric. Wilson endorsed Ferenczi's claim that tic formation was "facilitated by a need or desire to draw attention and feeling to certain parts of the body which belongs to the group of erotogenic zones and by their 'genitalization.'" Ferenczi's insight was confirmed by Wilson's clinical experience in which patients' development of successive tics served as a search for pleasurable sensations.[25] Again drawing on Brissaud and Meige and Feindel, Wilson emphasized the importance of habit: "Any purposive co-coordinated act passes by dint of *repetition* into a *habit* . . . Should the now purposeless habit assume an exaggerated form and a haphazard incidence, it has degenerated into a tic." Thus, Wilson insisted that movements that began as reactions to an irritation were often sufficient to set in motion the processes which resulted in a tic.[26] Even those tics that were caused by chorea could be transformed eventually into psychic habit tics.

Because as the case of O. demonstrated, the key to controlling tics rested with the patient's volition, strict management was required. "The association of fond parents and a spoilt child," Wilson reminded his audience, "is too frequent an antecedent to the development of a tic in the latter to be a mere coincidence. Sometimes one glance at the mother or father of the youthful tiqueur suffices to explain all." In such cases, patients have little chance of improvement and physicians must be firm in stipulating "a disciplinary régime calculated to recreate or develop self-control in matters not directly concerning his tic, thus exercising an indirect effect on it." Treatment might include use of sedatives, hypnosis, and especially Brissaud's "immobilization of movements," Meige and Feindel's "mirror drill," and Pitres's and Cruchet's "respiratory exercises."[27]

At the completion of Wilson's psychoanalytic explanation, René Cruchet briefly presented objections to some of Wilson's assertions.[28] As he had two decades earlier, Cruchet challenged Meige and Feindel's paradigm, including Wilson's attempt to update it with psychoanalytic assumptions. Although he admitted that many ticcers appeared to display abnormal mental states "characterized by mental infantilism, instability of ideas, inconstancy, carelessness, absence of will, impulsiveness, excessive emotivity, capricious affectivity, disorderly imagination, and loss of deliberation and judgment," many others were extremely intelligent, strong willed, and present with no evidence of hereditary mental defects.[29] Moreover, Cruchet did not believe that a habit illness was always present because in many cases the tics were executed independent of the patient's will. In these cases, Cruchet insisted, the ticcers were as much spectators of their tics as other observers. Rejecting psychological causes in favor of organic origins, Cruchet was convinced that the cause of many tics was cerebral, often connected with prior encephalitis.[30] "In cases where encephalitis provokes by its diffusion mental disorders and spasmodic reactions, it is very difficult to separate the mental from the spasmodic cause."[31] While he continued to believe in the breathing exercises that he and Pitres had developed for ticcers, Cruchet also advocated medications, including antipyrine [fever-reducing drug], potassium-bora-tartrate [drug to treat infection], gardenal [phenobarbital], and occasionally bromide [a sedative].[32]

Although most participants supported Wilson's psychoanalytic view, the discussion revealed the persistence of more than a few clinicians who were persuaded that neurological and infectious causes underlay tic behaviors. In an attempt to resolve this controversy, or at least provide guidelines for treatment for practitioners, in 1928 the *Lancet* solicited W. Russell Brain of London Hospital and the Hospital for Epilepsy and Paralysis to write an article on "The Treatment of Tic." Brain defined tic as "a coordinated purposive act, provoked in the first instance by some external cause or idea" that is "perpetuated as a repetitive involuntary movement." He pointed out that tic had recently also been used to describe involuntary movements caused by epidemic encephalitis. Brain argued, however, that it was misleading to confuse psychic tics with postencephalitic tics because the two had very different causes. Having differentiated organic from nonorganic causes, Brain dispensed with the possible organic substrate of convulsive tics by removing those movements with organic pathogenesis from the tic equation. "Tic," insisted Brain, properly should be attached to its "older and established usage of a derangement of neural function occurring in the absence of organic disease."[33]

Brain then described the origin of tic in much the same way that Wilson had done—by combining Brissaud's habit tic with a modified psychoanalytic characterization of child/parent conflict. "A child's innocent and physiologically justifiable sniff or wriggle may evoke admonitions or punishment from its father or the irritated protest of its mother that 'it gets on her nerves.'" As a result, Brain claimed, the child associated the movement with a combination of fears and resentments that manifested themselves as behaviors aimed at exasperating the parent. The child's reactions to the parent's scolding became linked to the original, but no longer present, stimulus. Now, the parent's admonitions or even the child's imagination of potential admonitions were capable of evoking the movements. Later on, others who the child viewed as acting like a parent might serve to provoke these movements. Thus, Brain explained, once the tic was established, the patient became powerless to control the tic. This set up "a vicious circle," reinforcing the feelings of inadequacy and self-pity that evoked the tic in the first instance. Given this causative psychological scenario, only psychotherapeutic treatment was likely to cure or control tics. Most important, "the whole household must unite in a conspiracy to take no notice of the child's movements, except as may be necessary to carry out the prescribed treatment."[34]

If the editors of the *Lancet* had solicited Brain's essay to help resolve the contradictions between the organic and psychogenic explanations for tics, they were disappointed. Practitioners seeking guidelines for treatment of patients must have been puzzled by the self-fulfilling definition that Brain used to separate organic from habit tics. As a result, the two views of tic seemed no closer to resolution at the end of the 1920s than they had been a century earlier. Rather than talking to one another, competing schools of explanation cited each other's work only within the terms of a medical convention, whereby competing claims were acknowledged in order to be written out of the discussion that followed. This recognition of alternative ways of knowing thus served as a literary device rather than as a sustained attempt to find common ground. Certainly, Brain's 1928 essay exemplifies this form of medical writing. By the decade's end, the rise of a new organic theory of disease causation, focal infection, illustrated why it was so difficult for physicians to transcend the competing (psychological versus organic) assumptions they held about the factors that led to ticcing behaviors. Believing in an underlying somatic cause, these physicians argued for a different course of treatment, one which emphasized surgery.

Focal Infection

In the teens and twenties the organic view of tics, as promulgated by Edwin Straus, received support from the emergence of what was called the focal theory of infection. Focal infection theory presumed that certain strains of bacteria appeared to infect specific targets, such as dental sockets, sinuses, tonsils, bronchial and pulmonary cavities, and the colon, leading to specific diseases like bacterial endocarditis, chorea, nephritis, appendicitis, and rheumatoid arthritis.[35] The chief exponent of focal infection was Frank Billings, professor of medicine at the University of Chicago and Rush medical schools and president of both the American Medical Association and the Association of American Physicians. Drawing on the laboratory findings of his colleague E. C. Rosenow, Billings outlined his arguments in a 1912 article in the *Archives of Internal Medicine* and in a series of lectures presented at Stanford University Medical School in 1915 and published in 1917.[36] Rosenow, who had isolated streptococcal bacteria from patients with some of these afflictions, reported in a 1914 *Journal of the American Medical*

Association article that these bacteria "injected intravenously in animals, tended to provoke inflammation in the same organs that had been affected in the patients from whom they had been obtained."[37] From this Rosenow had concluded that certain strains of streptococcus had an affinity for certain target organs.[38] Based on Rosenow's data, Billings claimed that removal of the infected organ site eliminated the disease.

Given Billings's credibility and stature, numerous American (and not a few European) physicians followed suit, performing a variety of minor and major surgical procedures for intractable infections from tooth extractions to tonsillectomies to hysterectomies. It did not take long before psychiatrists began to make a connection between focal infection and intractable psychiatric conditions.[39] Most influential among these was Henry Cotton, medical director of the New Jersey State Hospital in Trenton, who argued that infected sites produced bacterial toxins that migrated to the brain, causing an array of mental disorders.[40] Beginning in 1918, Cotton performed a variety of surgical procedures ranging from removal of previously filled teeth, tonsils, and sinuses, to major interventions like colon resections and hysterectomies on mentally ill patients. From 1919 to 1920 alone, Cotton performed two hundred such procedures; by 1922 Cotton had operated on 1,720 mental patients, claiming improvement in 77 percent.[41]

Influenced by Billings's authority and the publicity surrounding Cotton's claims, focal infection became a routine consideration for the treatment of Sydenham's and convulsive tics at a number of centers in the early 1920s. Typical was New York Hospital (Cornell Medical Center), where a review of all records of patients admitted from 1920 to 1930 for treatment for convulsive tics, habit spasms, and chorea (many of whom today would be diagnosed with Tourette syndrome) reveals that tonsillectomy was almost always performed, even when the patient had uninfected tonsils.[42] Even though tonsillectomies and removal of sinuses often failed to improve patients' tics and vocalizations, New York Hospital physicians continued to favor this procedure as a first line of intervention in the 1920s and early 1930s.[43]

By the decade's end, many practitioners agreed with Laurence Selling of the University of Oregon Medical School that focal infections were the primary causative factors in the production of tics. Selling built his case by combining Straus's arguments about the role of infection in convulsive tics with Billings's work implicating chorea as a focal infection disease and Cotton's claim that psychiatric disorders had a focal infectious etiology. In a paper delivered to the 1929 Annual Meeting of the American Neurological Association, Selling explicitly rejected the psychogenic origin of tics "laid down by Charcot and Brissaud, and Meige and Feindel," arguing that new evidence demonstrated the role of infection in the etiology of tics.[44] Like Straus and earlier observers, Selling argued that tics were merely one possible outcome of a variety of infections and that the severity of the tics was related to the persistence of a particular infection.

Selling, like so many before him, remarked on the relationship between tic and Sydenham's chorea. He saw the two as generally, but not always, separate disorders, which nevertheless resulted from a similar infectious cause. "The movements in the two diseases differ," but Selling reminded his readers that "one can pick out isolated movements in chorea identical with those of tic, and conversely in cases of tic some movements identical with chorea." Although he conceded that Sydenham's chorea, like rheumatic fever, was a self-limited disease with an often predictable course, he noted that there were also recurrences in some cases of chorea, suggesting that the distinction between Sydenham's and convulsive tic was often arbitrary. Citing Straus, Selling noted that convulsive tics might develop as residual manifestations of Sydenham's. The strongest proof of the infectious origin of the tic, according to Selling, was that encephalitis produced a variety of hyperkinetic movement sequels, including complex convulsive tics.[45]

If a variety of infections could cause tics, the best interventions would involve removing the causes of the infections.[46] Selling reproduced the case histories and treatments of three of his patients whose tics he claimed to have alleviated by removal of some or all of their infected sinuses.[47] All three cases were similar to Selling's first patient, a fourteen-year-old boy who, after a severe sinus infection at age six, developed a variety of rapid and jerky tics, including blinking, face twitching, eye movements, throat clearing, and sniffing. At age ten the boy's condition improved, but his tics returned following a prolonged sinus infection. Over the next several years, the patient contracted frequent and protracted colds and sinus infections. The pattern continued until 1928 when a severe cold resulted in antrum sinus infection followed by intense, almost violent, tics.[48] The connection between sinus infection and tic, Selling explained, was too direct to disregard, and Selling decided to surgically remove both antrum sinuses in September 1928. Eight months later Selling reported that "no one who knew the patient before or since could doubt the result." Before the operation the patient was always tense and ready to jump or turn in reaction to any stimulus; he was continuously restless, and his tics varied in intensity. After the operation, Selling asserted, the boy's tics were minimal and almost all other symptoms had disappeared. "To the casual observer now, the patient is a normal, but rather nervous boy."[49]

Selling's other two ticcing patients received similar interventions, although the third boy, also age fourteen, initially had only his tonsils and adenoids removed; but when this failed to have any impact, both his antrum sinuses were removed also. This too failed to halt the tics, and subsequently Selling removed both ethnoid sinuses. "Since this operation," Selling reported, "the patient has been showing slow but steady progress." In another briefly reported case, Selling removed the patient's tonsils and adenoids without affecting the tics, but subsequent operations on the antrums, ethnoids, and sphenoids (another sinus) ameliorated the symptoms.[50] Selling never published follow-up studies. Thus, the duration of remission of Selling's patients beyond the eight months reported on his lengthiest and first case is unknown.

By the mid-1930s focal infection was increasingly attacked and finally discredited as researchers proved unable to repeat Rosenow's laboratory experiments. The subsequent introduction of antibiotics made these procedures irrelevant even to their remaining advocates.[51] Tied as it was to focal infection, Selling's more general observations about infection and tic also became suspect. The more rigorous connection that Straus had made between prior infection and subsequent appearance of motor and vocal tics, however, persisted side by side, though not in concert or conversation, with psychoanalytic and other behavioral explanations.

The history of treatment of ticcing patients in this period underlines the extent to which investment in a theoretical construct allowed physicians to privilege certain symptoms and behaviors and to ignore others. It also reveals the degree to which physicians' beliefs about the efficacy of an intervention framed the way they understood their patients' symptoms and influenced their definition of what behaviors constituted a disease.

Each explanation—psychoanalytic, habit tic, infectious, and focal infection—especially when placed in the context of a selected patient narrative, was compelling to some extent, but it is apparent that alternative readings of patients' case histories could have resulted in other diagnoses, treatments, and outcomes. From this distance it is impossible to learn if the claims of cures existed more in the mind of the physician than in the life of the patient. Yet, given the importance of suggestion and of shared beliefs between patients and practitioners in any medical intervention, it may not be too far-fetched to suppose that many of the symptoms of many of the patients described in these clinical narratives actually did improve, at least in the short run. Having conceded that much, we should not be misled by claims that rested on extremely brief follow-ups.

The main contribution of these conflicting claims was therapeutic confusion. By the 1930s practitioners often seemed so uncertain about the causes of tics and involuntary vocalizations that they were, much like Saussure, persuaded that each symptom might have a separate and unrelated cause. It was only a small step for others to call into question the entire set of beliefs that tics were *ipso facto* evidence of mental or physical illness. By the mid-1930s developmental psychologists would challenge the belief that ticcing behaviors were necessarily abnormal in the first place.

The Disappearance
of Tic Illness

6

By the 1930s practitioners had been offered a series of confusing and contradictory theories about the causes of convulsive tics. Some researchers reinserted ticcing behaviors into a dynamic in which tics were one of a number of sequels to Sydenham's chorea, while others focused on the postencephalitic or other infectious origins. Still other authors attributed tics to psychological causes or to a mixture of psychological and organic factors. In an attempt to apply these conflicting claims, some physicians decided to assign a different underlying cause to each type of symptom *in the same patient.* In contrast to this medical model, developmental psychologists, observing students in classroom settings, discovered that tics were a common feature of normal childhood development. These tics developed into habitual tics in those with slowed maturation of motor controls. Although meant as a critique of psychopathological explanations, the developmental view was easily refolded into Freudian categories. By the end of the decade tic, as a separate category of disease, fell into decline because its symptoms were understood as one possible, but not exclusive, result of an underlying psychological conflict or a previous infection. When Gilles de la Tourette's name began to reappear in medical literature, first in occupied Europe in 1941, his explanations would be shorn of their organic and hereditary baggage and portrayed as if his arguments had been exclusively psychogenic.

*   *   *

On August 20, 1928, a fifteen-year-old male named Frank was admitted to the neurologic clinic of Chicago's Cook County Hospital. Frank could run and jump, yet, despite normal muscle control and good coordination, he could not walk. Whenever Frank attempted to walk, "he would make a step or two and then briskly lie down on his back and turn somersaults in quick succession." Frank would repeat these movements until he was exhausted. He also would stand up, sit down, press his fingers over his right earlobe or rub the back of both ears. Frank could walk when supported slightly by another's finger, but even then he walked abnormally, hopping on one foot. When sitting, Frank contorted his legs and head to one side. Every five to ten minutes Frank would inhale deeply, shaking his right forearm, as if shaking down a thermometer. When questioned about his behavior, Frank replied that he could not restrain himself from performing these acts.

A physical examination revealed that Frank was poorly nourished, pale, and had a large head with prominent ears. His head was covered with bruises and lacerations and his teeth were decayed. Frank's left arm and leg were considerably smaller and weaker than those on his right side. His upper and lower extremities were so flexible that he could twist his leg around his neck. A neurological examination revealed that Frank failed to sense pain.

Frank had contracted a number of childhood infections including measles, whooping cough, and pneumonia. He also had influenza during the postwar epidemic of encephalitis lethargica. Frank's involuntary movements first appeared at age seven (1920) after he was frightened by a large dog. Because of his previous influenza, Frank's movements were diagnosed as Sydenham's chorea. In 1925, at age ten, Frank was committed to Dixon State Hospital where he tested in the middle range of the moron group. While at Dixon, Frank suffered from throat congestion and enlarged tonsils. He was reported to have been extremely mean and untidy at school and at home. Frank also had been arrested several times for stealing. At times he became very angry and cursed and was reported to have been cruel to his siblings and to other children. The Dixon examination concluded that Frank "masturbates and is a sex pervert." Baffled by this history and range of symptoms, Frank's examiners diagnosed him as having left-sided atrophy (physical underdevelopment), sensory disturbances, abnormal attitudes, and unexplained movements, especially related to gait.

Frank's story was recorded in 1930 by neurology professor George B. Hassin and two of his neurology residents, Arthur Stenn and H. J. Burstein, at University of Illinois College of Medicine.[1] The group used Frank's symptoms to review the competing paradigmatic explanations of the etiology of tics.[2] Unable to see their patient's symptoms in combination as conforming with any one of these theories, Hassin and his colleagues separated each symptom from the other, attaching it to either infectious, psychogenic, habitual, or inherited causes. "There is," they wrote, "hardly any relationship between the three main groups of disorders exhibited by this patient." They were convinced that Frank's sensory disturbances resembled hysteria and were unrelated to his tics. They connected the weakness on Frank's left side to a developmental defect unrelated to Frank's bizarre walking, standing, and sitting behaviors. These were "gait or attitude tics," even though they were unlike the rapid muscle contractions usually associated with a diagnosis of convulsive tics. Frank's imitation of others' actions was, however, consistent with tic disease. Even here, Hassin and his colleagues were uncertain if the patient had actually seen the circus movements that they assumed his strange walking was imitating. They admitted that "if the conception of tic is to be limited to abrupt, quick, sudden" muscle movements, "then none of the anomalies presented by the patient are tics."[3]

By the 1930s neurologists like Hassin and his colleagues, who were conversant with the conflicting paradigms that purported to elucidate the causes and classification of tics, seemed more confused than enlightened by what they knew. They were even uncertain about the definition of what a tic was, not to mention what conditions may have undergirded it. Their patient's symptoms were indeed complex and confusing, but in ascribing each symptom to a separate cause, these neurologists revealed the extent to which competing claims had made it increasingly difficult to understand the causes and treat the symptoms of ticcing patients.[4]

A Category Disappears

If by the 1930s clinicians could find no common ground in the definition of what caused, let alone what constituted convulsive tics, the logical next step seemed to be elimination of the category altogether as useful for understanding and treating these symptoms.

After decades of discussions and endless pages in medical journals, Gilles de la Tourette's disease (convulsive tics) almost disappeared from medical literature in the 1930s.

Although it is possible to postulate an actual decline in ticcing symptoms, such a conclusion seems much less likely than the possibility that these behaviors were classified and treated under other recognized disease categories. One compelling piece of evidence in support of this view is provided by examination of the patients' admission and treatment records at the New York Hospital-Cornell Medical Center from the 1920s to mid-1930s. Although only one patient was diagnosed with convulsive tics in the 1920s and three others as "habit spasm" from 1921 to 1931, many of the hundreds of others diagnosed with chorea presented with symptoms that both then and now would have fit the category of Tourette syndrome.[5] For instance, of the forty-five cases listed as chorea from 1926 to 1931, between 20 to 30 percent would, at other times, have been classified as convulsive tics.[6]

A nine-year-old boy diagnosed with "psychogenic chorea" in 1927 had displayed hand, head, and tongue movements for over a year, symptoms that his doctors believed were "aggravated by excitement." The movements progressively worsened and three weeks after admission an attending physician wrote that the patient's "twitchings very likely have a psychogenic nature but the determining factor [has] not [been] determined and [I am] of the opinion that this can only be done with prolonged residence and study in a carefully supervised environment not at home." The child's parents disagreed and the patient was released with a note that he had "lots of purposeless movements—getting worse."[7] A ten-and-a-half-year-old girl, who had recurrent facial tics and twitching for four years, was diagnosed as suffering from chorea even after she began to make involuntary noises.[8] Another "choreic" patient, admitted three times between 1926 and 1931, had head, neck, left arm, and left leg twitchings for four years prior to his first admission.[9] The physician who evaluated the twelve-year-old boy in 1928 believed that the patient "has been too carefully supervised by an anxious mother . . . He would be better off on leaving the hospital in a well supervised environment than at home. He displays mild neurotic symptoms which would probably be checked by proper supervision."[10] By 1931 the boy's purposeless head, face, hand, and right arm movements had intensified. Yet, after more than a decade of waxing and waning motor and vocal tics, the diagnosis remained Sydenham's chorea.[11]

Another piece of admittedly impressionistic evidence in support of the view that convulsive tics had become subsumed under other diagnostic categories is found in an examination of the *Index Catalogue of the Surgeon General's Office,* the official bibliographical medical publication of the United States government. The Second Series, published in 1913, listed more than 300 articles on tic, covering five double-columned pages.[12] Twenty years later, the Third Series, published in 1932, listed only forty citations under "tic."[13] By the time of the publication of the Fourth Series, issued in 1941 (which covered the 1930s), physicians looking under the heading "Gilles de la Tourette's Disease" were instructed to "see neurosis."[14] The "N" and "T" volumes never appeared because publication was suspended during the Second World War. However, in the American Medical Association's *Quarterly Cumulative Index Medicus* for 1938, one finds only five listings under "tic," two of which are on the topic of Parkinsonian tics.[15] When the Fifth Series and final series of the *Index Catalogue of the Surgeon General's Office* appeared in the 1950s and early 1960s, there were only three listings under "tic," the latest published in 1942.[16] None of the three viewed tic in the context of Gilles de la Tourette's typology.

In contrast, the diagnoses under the categories of neurosis and chorea grew exponentially, including many cases that previously would have been classified as tic. For instance, the *Quarterly Cumulative Index Medicus* had listed only thirty citations of chorea for 1930–1931,[17] but by 1938, in the same series of *Index Catalogue of the Surgeon General's Office* in which readers searching for Gilles de la Tourette's Disease were instructed to search under "neurosis," there were ten double-columned pages listing almost one thousand citations on the topic of chorea, including a new category labeled the "mental manifestations" of Sydenham's chorea.[18]

From 1930 through the end of 1939, fewer than forty articles were published in medical journals in all European languages— none appeared in other languages—on all forms of tics, including those seen in patients diagnosed with Parkinson's disease. Only one of the essays, George Hassin's, viewed Gilles de la Tourette's typology as diagnostically useful. This declining number of articles does not *prove* absolutely that many symptoms and behaviors had been reclassified as parts of other conditions, but close reading of studies published in the 1930s makes it difficult to arrive at an alternative explanation. It was as if the debates over the causes of motor and vocal tics in the 1920s appeared to have so exasperated researchers by the 1930s that they gave up trying to explain the phenomenon in terms of its symptoms. If the symptoms were merely one possible outcome of an underlying infection or psychological conflict, then labels such as maladie des tics, Gilles de la Tourette's disease, or convulsive tics would lose their clinical utility.

Finding a Sequel to Infection

Throughout Europe in the 1930s neurologists and psychiatrists sought to balance the growing view that tics were psychological with the increasingly persuasive evidence that both motor and vocal tics often were sequels to encephalitic infections. In these discussions, there was rarely, if ever, even a mention of Gilles de la Tourette or Charcot. For instance, in July 1930 Generoso Colucci, Director of the Institute of Experimental Psychology at the University of Naples, published a report citing the cases of four ticcing patients as examples of how symptoms that appeared to be psychogenic or hysterical were, on further examination, very often sequels to a previous encephalitic infection. Using photographs of his patients supplemented with encephalagrams, Calucci contrasted what he believed to be the subtle, but distinct, presentations in those with postencephalitic tics from those whose tics were psychogenic.[19]

Like Calucci, other clinicians who viewed tics as a sequel to encephalitic infection found no reason to refer to Gilles de la Tourette or his diagnostic categories.[20] At the annual 1936 meeting of the Société Medico-Psychologique, French psychiatrist G. Heuyer presented a case in which only one of a pair of identical twins developed convulsive tics. Although both girls had inherited their father's epilepsy, only the ticcing twin had contracted encephalitis. This, argued Heuyer, was persuasive evidence that convulsive tic, rather than a hereditary disorder, was a "post-encephalitic neuro-psychiatric syndrome."[21] The panel's three commentators agreed with Heuyer's assessment that the girl's tics were postinfectious. One of the commentators added, however, that in "certain sequels attributed since 1920 to epidemic encephalitis" tics could also be produced by "hereditary-syphilis," which produces symptoms that are encephalitic because syphilis also damages the brain in the same locations as an encephalitic virus.[22]

This connection between tics and syphilis was widely discussed in the 1930s. By the decade's end French physician Lucien Cornil reported that sixteen out of thirty-one ticcing patients tested positively (via the Wassermann, Bauer-Hecht serological tests) for syphilis, suggesting that there was *"a particular frequency among ticcers of signs of presumption or certitude of hereditary syphilis."*[23] Citing the evidence that both encephalitis and Sydenham's chorea resulted in tics because both types of infection produced lesions in subcortical areas of the brain, Cornil argued that other conditions, such as syphilis, which also produced similar brain lesions, logically could cause tics. Anything that causes lesions in the grey matter of the brain could, according to Cornil, produce involuntary movements, including tics. Such a view, he concluded, demands "a fundamental revision of the pathogenesis of tics."[24]

In his discussion of brain lesions and tics Cornil referred to a 1935 study authored by Mildred Creak, chief of the Children's Department and two of her colleagues at Maudsley Hospital in London.[25] In that paper Creak and her colleagues specifically rejected the view that either tics in general or Gilles de la Tourette's disease in particular should be viewed as a separate disorder from chorea. "It is now generally held," they wrote, "that tics, even the *maladie des tics* of Gilles de la Tourette, do not constitute a disease entity."[26] Influenced by the work of the German Edwin Straus and others who believed that encephalitis caused lesions in the basal ganglia of the brain,[27] Creak concluded that motor and vocal tics displayed by six of her patients were sequela to chorea. Moreover, relying on the systematic studies of heredity and chorea in families,[28] she argued that postchoreic movement disturbances, including involuntary vocalizations, were hereditary. Tics appeared only in that subset of choreic patients with a familial predisposition for tics. This connection between tic and Sydenham's chorea, Creak explained, "does not decrease the importance of constitutional factors, even though chorea may be an infective disease." She also rejected Meige and Feindel's claim that imitation and habit played important roles in the genesis of tics. But she was ambivalent about other psychological factors because emotional stress appeared to increase symptoms. Although Creak was unwilling to endorse the claim that the cause of tics was purely psychogenic, she was willing to concede "that the neurosis chooses the organ (using that term in the Freudian sense) on account of the motor lability, acquired or given by constitution."[29] Nevertheless, by the mid-1930s physicians like Creak and her British colleagues were able to diagnose and understand the etiology and course of motor and vocal tics, even when accompanied by coprolalia and echolalia, without having to rely on Gilles de la Tourette's and Charcot's separate disease category, maladie des tics.

By 1936 French neurologist P. Guiraud could argue that each of the components that Gilles de la Tourette had linked in his typology—involuntary motor movements, echolalia, echopraxia, and coprolalia—could have a different possible neurological, psychiatric, or infectious cause.[30] Like Hassin and his American colleagues in 1930, Guiraud found that the echolalia or motor tics exhibited by one patient might have an entirely different set of underlying organic causes than those displayed by another.[31] Reflecting the neuropsychiatry of his day, Guiraud was a "splitter." Thus, while he connected a number of his patients' motor and vocal tics to schizophrenia, Guiraud saw schizophrenia itself as a set of different disorders, with each set having distinctly different underlying causes.[32] This type of thinking was reflected in a new category of "mental manifestations" of chorea that appeared in the Fourth Series of the *Index Catalogue of the Surgeon General's Office*. As a result, an increasing number of studies on chorea and infection examined patient behaviors that earlier would have been attached solely to convulsive tics.[33]

Guiraud's position reinforced the advice about treatment of tics given in 1932 by the influential and eclectic American psychiatrist Tom A. Williams.[34] Reviewing the cases of a number of patients, each of whom allegedly had different underlying causes for their tics, Williams concluded that it was impossible "as a rule" to treat the tic itself because a tic was merely a symptom of a more complex underlying disturbance. Therefore, he insisted that whether the underlying cause of the tic was "a neurologic lesion or whether it is a chemical disturbance, or whether it is a reaction to circumstance, purely psychic, the etiological factor, broadly speaking, must be ascertained" before treatment can be attempted.[35] If, as Williams believed, there were multiple and different causes of tics, Gilles de la Tourette's typology had no diagnostic utility.

Developmental Psychology Emphasizes Normal Behaviors

If neurologists like Williams and Guiraud subsumed ticcing behaviors into subsets of other organic disorders, the new, mainly North American discipline of developmental psychology viewed convulsive tic symptoms as a form of normal early childhood behavior distorted by environmental pressures. Led by academic psychologists, developmental approaches relied on the collection and classification of survey data using social science models. Emphasizing that tics were a common feature of normal childhood, developmental studies dispensed with classifications based on specific tic behaviors such as that elaborated by Charcot and Gilles de la Tourette.

Most representative of the developmental approach was *A Study of Tics in Pre-School Children* published in 1935 by University of Toronto researchers William Emet Blatz and Mabel Crews Ringland.[36] Based on their observations and tests of seventy-one normal schoolchildren—twenty-five nursery-school children and forty-six kindergarten and first grade pupils, Blatz and Ringland reported that their study confirmed earlier observations that tics were a common form of behavior in normal preschool children.[37] They endorsed the views of those psychologists "who repudiated the claims of a sexual basis for tics on the ground that this was both illogical and inconceivable," and were skeptical about the claims that tics resulted from a "neuropathic constitution."[38] Along with the American developmental psychologist Mary Chadwick, Blatz and Ringland believed that tics only developed into a behavioral problem when a child's movements were inhibited. They subscribed to Chadwick's view, which appeared in her influential 1928 text, *Difficulties in Childhood Development,* that tics were "symptoms of unhappiness in a child's life for which some adult is responsible." The blame rests on "errors in training, plus the neurotic character-traits of those in charge of young children."[39]

Blatz and Ringland also agreed with fellow developmentalist Annie Dolman Inskeep, whose 1930 study, *Child Adjustment in Relation to Growth and Development,* claimed that "habit spasms, or tics, originally had a purpose and are not made when the muscles are being used for something else." This explained why children with habit spasms adopt one tic and then drop it, replacing it with another. Uncomfortable clothing, reported Inskeep, frequently caused habit tics "in the need to overcome discomfort, as hitching up trousers or a lingerie strap, or throwing back a lock of hair." Because these habits often appeared at puberty, they affected accessory muscles such as those controlling the eyes or mouth. According to Inskeep, habit tics were due to slowed physical maturation, when a child's motor skills developed more slowly than normal. Most tics, however, could be traced to the demands of modern schooling. Ill-fitting desks exacerbated pressures on the child whose growing body was not adapted to sitting still for long periods of time. For Inskeep "one of the most cruel punishments . . . is to give a child nothing to do and force it to sit still."[40]

Blatz and Ringland's study confirmed the work of Chadwick and of Inskeep that "force and punishment are worse than useless in modifying this type of behaviour." By recognizing and testing for the role of social restraint, the authors believed their study had explained why, regardless of the group, tics tended to increase with age. Emphasizing tics as normal, rather than pathological, Blatz and Ringland found that contrary to most observers of tic, there was no significant difference between girls and boys. Tics, they concluded, resulted from a conflict between children's desires with the requirements of their environment. "Left to himself a child is in a state of constant motor activity, but when gross bodily movements are inhibited, even temporarily, a conflict is created and this active impulse overflows into a great range of fine muscular movements we call tics."[41]

After this brief but intense flurry of interest in the 1930s, developmental psychology disappeared from the discussion of tics. Perhaps this was because so much of what developmental psychologists had to say about child development and restraint could be placed in psychopathological psychoanalytic categories, despite the attempts of developmental psychologists to emphasize the normal. Moreover, developmental psychology flourished mainly in North American university psychology departments, while in continental Europe by the 1930s child psychology had fallen under the sway of psychoanalysts like Anna Freud and her rival Melanie Klein, both of whom, as a result of Nazi policies, migrated to Britain by the end of the 1930s. They and their disciples trained a new generation of child analysts including Joseph Wilder, Margaret Mahler, Ernst Kriss, and Erik Homberger (later Erikson), most of whom, again because of the Nazis, settled in the United States. As a group these migrants would unite (albeit sometimes after some tension) with North American psychoanalytic psychiatrists and together would overwhelm developmental psychology in Canada and the United States.

Psychoanalytic Explanations Persist

The primary challenge to the psychoanalytic frame in the 1930s remained the reports by organic-oriented psychiatrists and neurologists who offered organic explanations that tics were a sequel to infectious encephalitis and acute chorea. This issue was addressed directly by psychiatrist Curt Boenheim of the Polyclinic for Nervous and Difficult Children at the Childhood Disease House of the City of Berlin. Boenheim reported that an examination of ninety-one cases, of which forty-nine had been diagnosed with tic and forty-two had been diagnosed with acute chorea, revealed that those with tic, but not chorea, had a significantly greater family and personal history of mental disorders than those with chorea. Thus, Boenheim concluded that although there were rare "borderline" cases of tic combined with chorea, the two conditions were distinct disorders with different underlying causes.[42] Having made the distinction, Boenheim argued that compulsive tics were enabled by repressed childhood sexual conflict. The key issue, he insisted, was not the underlying cause of tic, which had been the focus of the debate of the last decade, but whether or not psychological conflicts facilitated these constitutional and environmental features to develop into a "psychotic or neurotic compulsive phenomenon."[43]

Boenheim's claims, like those of his organic and developmental antagonists, created a context in which tics would not be classified as Gilles de la Tourette's disease. Medical textbooks published in the 1930s provide further evidence that tics were now fully subsumed under other categories. As a result, clinicians trained in American medical schools learned that tics were always symptoms of either chorea or, if no prior rheumatic fever was uncovered, a psychogenic illness that required unraveling of repressed psychodynamic issues. Certainly, that is what they learned from a typical text, *Common Neuroses of Children and Adults* (1937), written by two Temple University Medical School professors, Oliver Spurgeon English and Gerald H. J. Pearson.[44]

Acknowledging the need first to eliminate a possible diagnosis of chorea, English and Pearson focused on the psychogenic, where tics always represent "an expression of some erotic desire, either the desire for physical gratification of a need for love or affection, or for autoerotic pleasure. Usually this is accompanied by a desire to express some aggressive hostile wish," which is "accompanied by a fear of punishment." These wishes and fantasies were also expressed by the type of movement. For instance, eye blinks could be read as symbolic of masturbation. Because the child learned early on that a movement attracted attention, the child was willing to tolerate ridicule and scorn, making tics difficult to treat.[45]

The authors presented several typical cases of tic. What is most interesting about these cases is that their psychoanalytic frame is never justified or even authorized by citation of a psychoanalytic authority. Rather, the cases and conclusions are presented as if clinicians reading them shared a common understanding of human behavior rooted in Freudian categories without relying on a technical Freudian vocabulary.

Discussing the case of a ten-year-old boy with rhythmic blinking, English and Pearson reported that the patient was obsessed by a fear that his eyes would be put out. The boy's initial tic started soon after an eye injury that resulted from the mother's hitting the boy while punishing him for masturbating. According to the authors, the patient was extremely curious about girls' bodies and "through this desire to peep he had observed parental intercourse on a number of occasions." The boy's eye blinks thus were connected to his desire to gain sexual stimulation by peeping, but what he had seen had been repulsive and frightening. "He had seen his sister's and mother's genitals, compared them with his own," and "wondered how they [his sister's and mother's] became so ugly and mutilated and feared that something similar would happen to his own if he continued to have sexual feelings and masturbate." These fears were intensified, according to English and Pearson, by his parents' threats "that if he continued to masturbate his penis would be cut off." Because the boy often had observed his parents quarreling, "he believed his father was cruelly hurting his mother's genitals with his penis," a fear that was reinforced when soon after, his mother went to the hospital to give birth to the boy's younger brother. Both of these events increased his fears of being abandoned by his mother. Because the boy's "eyes had been the avenues through which the sexual stimulation and the associated unpleasant ideas had come, he endeavored to be blind to the whole situation and whenever he was sexually stimulated or had the desire to peep he tried to fulfill the Biblical injunction, 'If thine eye offend thee, pluck it out'; i.e., he blinked as if he had no wish to see, and as if he tried to protect himself from blinding as a punishment for seeing."[46]

English and Pearson presented several similar examples, noting that in all cases it was evident that the movement was "an attempt to express a sexual wish or idea." Because these involuntary movements represented the dramatization of a conflict between instinctual desires and inhibitions, therapy directed at the symptom, such as habit training, would fail.[47] If the tics themselves were not to be the focus of treatment, but rather signs to be read on the way to uncovering repressed psychosocial conflict, it is not surprising that clinicians trained in American medical schools in the 1930s and 1940s were unfamiliar with Gilles de la Tourette's illness.

The War Years

The outbreak of the Second World War and the Nazi policies preceding it profoundly influenced medical thinking about convulsive tics in a variety of ways. Psychoanalysis was ripped out of its central European home and transported almost in its entirety to Britain, the United States, and Canada. Those nonpsychoanalytic psychiatrists who remained, a number of whom had advocated the organic nature of movement disorders, were isolated during the war. After the war their work was viewed with the suspicion that it reflected the influence of Nazi racial assumptions. Replanted in the English-speaking United States, Canada, and Britain, psychoanalysis thrived during the war and overwhelmed all other ways of understanding vocal and motor tic symptoms in the postwar world.

As a result of these factors, work on tics that took place during the war in either German or German-occupied countries has been almost totally ignored, although the ideas examined in some of these studies were elaborated by German researchers during the 1950s. One of these studies, "Anatomical Clinical Studies of Complex Hyperkinetic Syndromes," published in 1941 in a Berlin journal, called for a reinstitution of Gilles de la Tourette's typology. Because of where and when it originated, it had no immediate

influence on the subsequent views of those concerned with tic behaviors. Its authors, psychiatrist Andre DeWulf and neurophysi-
ologist Ludo van Bogaert, living in occupied Belgium, argued that
the classification and etiology of motor and vocal tics described by
Gilles de la Tourette remained valid. DeWulf and van Bogaert based
their conclusions on an autopsy of a thirty-year-old man, Edgard
W., whose behavior, they claimed, "reflected, trait by trait, the
affliction described by Gilles de la Tourette."[48] Relying on Edgard's
case, they ruled out both an organic or infectious etiology, claiming
instead that Edgard's life history validated Charcot and Gilles de
la Tourette's classification of convulsive tics as a separate disorder.

DeWulf and van Bogaert reproduced verbatim an entire section
of Gilles de la Tourette's 1899 revised view emphasizing the psychological aspects of convulsive tics,[49] rather than the 1885 study,
which concluded with hereditarian claims.[50] They added that following Gilles de la Tourette's initial description of the disease, "the
published cases were abundant," but soon after Meige and Feindel
published their study, "the illness disappeared," so that it "has become extremely rare, these last years, or at least, rarely mentioned
at the meetings of neurologists and psychiatrists." As a result, the
authors found no cases in the literature "which had been examined
with modern histo-pathological methods."[51]

DeWulf and van Bogaert presented an extensive case history of
Edgard W. along with six photos of him, naked, showing the various positions caused by his motor tics and vocalizations. The photos were actually frames from a film that had been made in 1937
in the months before Edgard's death from what appeared to have
been a severe epileptic type of convulsion. Edgard's symptoms—
shouts and tics—first had appeared twelve years earlier when he
was eighteen years old and institutionalized. The symptoms included convulsive mouth and face movements and involuntary tics
involving his left hand and leg. In the course of these movements,
Edgard would shout out an inarticulate word that sounded like
"goo," which sometimes was followed by the word "pig." These
symptoms had become much stronger over the years, making
Edgard unable to function in society.

Edgard was transferred to Corbeeck-Loo Institute (an asylum)
in 1934 where DeWulf and van Bogaert followed his case. His
physicians uncovered no neurological or infectious factors that
could explain the tics and vocalizations. Their report focused on

the young man's mental condition. DeWulf and van Bogaert described the patient as lucid, with a trustworthy memory and sane judgment. Edgard was reported to be slightly irritable with the asylum personnel and the other patients, but DeWulf and van Bogaert found that understandable given his affliction. "There isn't any depression. The patient, to the contrary, fights courageously against his illness." However, they depicted a lonely and sad existence for Edgard. For instance, Edgard rationalized that his family rarely visited him because of the expense of the trip and "the repellent aspect of his illness." Edgard remarked that "he should be condemned and only wanted to die quickly." Yet, DeWulf and van Bogaert claimed that Edgard "however had no melancholic or suicidal ideation."[52]

In the two years before his death, Edgard was injected twice daily with a combination of scopolamine,[53] morphine, and the sedative phenobarbital. As a result, mainly of the phenobarbital, Edgard experienced mental confusion and painful contractions of his hands for which there was no relief. Three weeks before his death Edgard complained "that no one left him in peace even at night, that he was being pricked with needles and that he had not been given anything to eat. His general condition is therefore very bad." In February 1937, thirty-year-old Edgard suffered an extremely violent convulsive fit while taking a bath and died.[54] Edgard's medications had killed him.

DeWulf and van Bogaert's autopsy on Edgard, performed four hours after his death, included a meticulous microscopic brain cell examination and a complete brain and spinal cord dissection. Aside from "some cellular sclerosis" in the "frontal, parietal, and motor regions" of the brain, DeWulf and van Bogaert wrote that they had "uncovered no other detail which merited being mentioned. These cellular lesions have no significance given the general ill health of the patient and the toxic medications which were administered to him during the final weeks of his life." Of course, even in 1941 many other neurologists and neurologically informed psychiatrists would have viewed the sclerosis in the frontal, parietal, and motor regions as evidence of an organic cause of Edgard's motor and vocal tics. Based on Edgard's autopsy, however, DeWulf and van Bogaert concluded that this "clinical description of a typical case of tic disease (Gilles de la Tourette)" contradicts the claims of "certain authorities [who] have tried to represent the tic disease as a result of an infectious illness of the central nervous system, especially of the encephalitis epidemic." They also ruled out any connection with chronic chorea or brain cell degeneration.[55]

Eight years later, in 1949, van Bogaert reviewed his and DeWulf's 1941 conclusions, including the pictures of the naked Edgard, in the French (Paris) medical journal *Traité de Médecine*. There was no reason, wrote van Bogaert, to modify Gilles de la Tourette's 1899 description of this disease or to reconsider the conclusion that he and DeWulf had reached in 1941.[56]

Van Bogaert's 1949 revalidation of Gilles de la Tourette's typology was unique, although the earlier treatment of Edgard with lethal experimental interventions was, unfortunately, not exceptional. But to the extent that he had characterized and endorsed Gilles de la Tourette's view of motor and vocal tics as exclusively psychological, van Bogaert's claims were compatible with the overwhelmingly dominant medical assumptions of the 1940s. Nevertheless, when it appeared in 1949, van Bogaert's reliance on Gilles de la Tourette for a psychogenic theory must have seemed quaint even to those who rejected an organic explanation of the etiology of tics. By this time psychogenic explanations were synonymous with psychoanalytic claims emanating from North American psychiatry.

The Second World War resulted in the United States' domination not only in postwar economics and politics, but also in medicine. The Nazi expulsion (and murder) of psychoanalysts of Central Europe had forced survivors to regroup on Anglo-American soil. After the war American psychiatry, like the American economy and military, emerged unchallenged. Dominated as they were by psychoanalysts, when American psychiatrists attacked the issue of tics, they took—in terms of other competing points of view—no prisoners. By the mid-1940s, when psychoanalysts appropriated Gilles de la Tourette's name for a symptom description, they simultaneously gobbled up pieces of developmental theory, retranscribing it into psychopathological language. Like much of the American empire that emerged after the war, American psychoanalytic psychiatry had used the war years for postwar planning.[57] This psychiatric community, including its now powerful emigré psychoanalysts, would hitch a ride on the nation's military and economic machine and exert a powerful influence on postwar medical practice well beyond its borders. Psychoanalysts resurrected Gilles de la Tourette's name, attaching his typology to a variety of psychoanalytic categories of explanation. There would be no room in the psychoanalytic tent for the tic *disease* as described by Charcot's extern, however. Rather, under the guidance of Hungarian-born, Viennese-trained Margaret Mahler, a new generation of American psychiatrists would learn that the symptoms described by Gilles de la Tourette were signs of a deeper psychosexual disturbance, albeit informed by organic factors. Mahler and her American colleagues would transform Gilles de la Tourette's tic disease into a syndrome, in which the tic itself would be less the object of treatment than some putative, repressed set of childhood sexual conflicts.

Margaret Mahler
and the Tic Syndrome

$$7$$

Ever since Freud first conceded that the origin of convulsive tics was most likely organic, psychoanalysts struggled to interpret the master's assertion in a way that authorized psychoanalytic intervention. Thus, Sandor Ferenczi reread Freud's remark in the context of a theory of narcissism, concluding that tic appeared in those with a *"constitutional narcissism, where the smallest injury to a part of the body strikes the whole ego."*[1] This logic dictated a psychoanalytic view of tic that dispensed with the need to consider the nature or extent of any organic features except as physical manifestations of an underlying psychosexual disturbance.

Ferenczi's interpretation, especially as it was neither challenged nor endorsed by Freud, provided most psychoanalysts with the justification they needed to focus on the psychogenic aspects of convulsive tics. However, by the mid-1930s such a strategy seemed more difficult in light of increasing evidence of the role of encephalitis in the production of tics. Added to this was the persistent connection between acute chorea and a variety of involuntary movements. That both encephalitis and Sydenham's chorea appeared to act directly on brain structures made it much more difficult for psychoanalysts to ignore an underlying organic cause of convulsive tics.

Psychoanalysts had met these challenges with several strategies, including classifying psychogenic tics as separate from those with identifiable infectious features.[2] Psychiatrists would continue to pursue such explanations, which were not unreasonable given the general state of neurological science in the late 1930s.[3] Some psychoanalysts, however, refined their explanations of the interaction of psychic and organic spheres in a way that appeared to account for the increasing evidence that tics had an organic substrate and simultaneously justified psychoanalytic interventions as the most effective clinical approach. This became known as psychosomatic medicine. By the late 1950s psychosomatic approaches would be attached to a number of emotional disorders that earlier psychoanalysts had viewed only in a classical psychogenic frame, including motor tics and involuntary cursing.

Early Psychosomatic Theories

When ruling out any physiopathology in the autopsy of a patient afflicted with convulsive tics, Ludo van Bogaert had cited the prewar writings of the Viennese psychoanalyst Joseph Wilder.[4] Van Bogaert had found Wilder's work compatible because it developed an elaborate theory that explained why convulsive tics so often appeared to be organic.[5] According to Wilder, convulsive tics were "a classical example of a psychosomatic condition" in that they had an organic impact and manifestation, but their persistence and evolution were due to a set of underlying psychological factors.[6] The "psychic" part of the equation included "some sort of consciousness, clear, dim, or subconscious," while the "somatic" elements were "those phenomena in which we cannot detect any trace or form of consciousness." Conscious actions (for instance, as in gum chewing), Wilder explained, often became automatic functions and thus appeared to be under organic control. Therefore, the "conscious" or "psychic" aspects of tics became automatic as their psychological origin was obscured by their seeming involuntary action.[7]

Although Wilder conceded that a few tics were the result of purely organic illnesses, especially those of postencephalitic origin, even these "minority of tics" were sometimes "psychogenic" because the organic damage resulted in the release of unconscious mechanisms, resulting in exciting the brain mechanisms responsible for motility. But because most tics were purely psychogenic, it was the psychological situation that created the "need for symptoms." Those with insufficient training in self-restraint and immature ideation and volition were particularly prone to regress partially or totally to unconscious automatic infantile patterns of behavior.[8]

To be brought under control, the psychological causes of tic had to be made conscious once again. However, because the tic had become somatic (automatic), it was, Wilder warned, difficult to eradicate. Wilder conceded that psychogenic tics often remained after a successful psychoanalysis or appeared for the first time during analysis. Thus, treatment of tics usually required "not only the realization and removal of the psychic regression leading to tics but also a difficult training process which may be called *deautomatization.*" This was never easy, Wilder pointed out, because patients had to be made conscious of movements that had become completely automatic. Once conscious of their tics, these patients had to develop the willpower to restrain other stimulations from resulting in new tics.[9]

The Construction of Tic Syndrome

Like Wilder, many Central European psychoanalysts fled in the wake of Nazism to the United States, where they joined forces with an increasing number of native-born American psychoanalysts. Among this group of emigré psychoanalysts was Hungarian-born pediatrician Margaret Schoenberger Mahler (1897–1985), who would refine the psychoanalytic view of tics, updating Ferenczi's explanations for a new generation of psychiatrists. While still in her teens, Mahler was "adopted" into the family of Budapest psychoanalyst Vilma Kovács, former student and analysand of Sandor Ferenczi. As Mahler would later write, Kovács became her idealized mother substitute and, as an influential female psychoanalyst, a role model for the younger, aspiring physician. The Kovács household served as a salon for Budapest's most influential psychoanalysts, including Ferenczi, Kovács, Michael Balint, Geza Roheim, and others. All of these Budapest analysts would play important mentoring roles in Mahler's career, but none more important than Ferenczi. After Mahler graduated from medical school, it was Ferenczi who urged her to consider psychoanalytic training. Under Ferenczi's sponsorship, Mahler entered a training analysis in Vienna in 1926, first under Helene Deutsch, and then with August Aichorn and Willi Hoffer.[10]

This was the period when Ferenczi laid out the classic psychoanalytic account of tics. In 1925, Kovács, following Ferenczi's lead, published an article that attributed the convulsive tics of a forty-five-year-old woman to early childhood "narcissistic" conflicts.[11] Thus, Mahler's formative professional years had been shaped by the most important psychoanalytic interpreters of tics.

Mahler arrived in the United States in late 1938, and after a brief association with Mt. Sinai Hospital, moved in 1941 to the Columbia University's New York State Psychiatric Institute and Hospital in Manhattan. From 1943 to 1949 Mahler produced a series of articles that reintroduced the designation "Gilles de la Tourette's Disease," remolding it into a psychoanalytic frame as "tic syndrome." In a 1943 article entitled "A Psychosomatic Study of Maladie des Tics (Gilles de la Tourette's Disease)," Mahler and neurologist colleague Leo Rangell described "the classical syndrome" as usually appearing at an early age as a series of abnormal, uncontrollable involuntary movements, which gradually increased in intensity and frequency. The body movements, reported Mahler, often were accompanied by involuntary utterances and inarticulate cries, which were followed by imitative behaviors (echolalia) and eruptive swearing (coprolalia).[12]

Although Mahler adopted Gilles de la Tourette's description of the symptom course, she parted company with his explanation of the underlying causes of these behaviors. If the symptoms of Tourette's disease were identifiable, the cause of the condition, wrote Mahler, was obscure. In contrast to Wilder, Mahler believed that convulsive tics most often had an organic substrate, but that tic behaviors manifested themselves only in those susceptible children who had experienced severe, repressed familial psychological conflicts. That is, the organic factors were necessary, but not sufficient by themselves to produce tics. "There is quite likely a substratum of organic disease," wrote Mahler and Rangell in 1943. "The important fact, however, is that this factor in itself would be insufficient to bring on the syndrome, but that it renders the individual defenseless against overwhelming emotional and psychodynamic forces." Therefore, like Wilder and Ferenczi, Mahler emphasized the need for intervention on the psychic rather than the somatic level.[13]

Mahler also cited the work of those who believed that the ticcing behaviors were caused by brain lesions, but emphasized that the exact mechanisms were unknown.[14] She excluded from tic syn-

Figure 7.1 Hungarian-born pediatrician and psychoanalyst Margaret S. Mahler immigrated to the United States in 1938, and from 1943–1949 produced a series of influential articles on "Gilles de la Tourette's Disease," updating Ferenczi's explanations for a new generation of psychiatrists.

drome those conditions, such as choreas, encephalitis, or schizophrenic-like psychosis, where the physiological mechanisms appeared to be clear.[15] Insisting that those with tic syndrome had no infectious substrate, Mahler focused on children with well-defined motor neuroses as described by Trousseau, Gilles de la Tourette, and Guinon.[16] Glossing over Guinon's rejection of Gilles de la Tourette's typology, Mahler presented past investigations as if there had been an agreed-upon and stable definition, which conformed exactly to her wish to separate one group of ticcing children from another. "We designate as patients with tic syndrome," wrote Mahler and her two residents, Jean Luke and Wilburta Daltroll, in 1945, "those children whose tics conform to *the classical accepted definition:* involuntary, lightning-like repetitious jerks of a physiological group or groups of muscles, ordinarily of the face, the neck, the extremities, etc."[17]

These symptomatic and etiological assumptions informed Mahler's diagnosis and treatment of ticcing patients. They reinforced her view that ticcers were "highly narcissistic individuals, who invest an undue amount of attention in their own bodies and who are unable to retain stimuli or irritations without immediate defensive enervation."[18] Mahler's clinical case histories served as models for a generation of practitioners. Psychoanalyst Louise J. Kaplan remembered in 1976 that "Mahler's 1943 [her first] paper on the tic syndrome was instantly recognized as a classic."[19] Coauthored with Rangell, who focused on the neurological issues, the paper discussed the ticcing behaviors of an eleven-year-old boy named Freddie. Freddie's case became the focus for discussion of a number of other psychoanalytic studies and an exemplar for psychosomatic tic syndrome.

Freddie was seven years old when his first eye blinking began. This was soon followed by head twitching, arm shaking, lip puckering, and tongue protrusions. At night, Freddie's body shook as he attempted to fall asleep. His mother often had to take him out walking in the middle of the night to relieve his restlessness. Then Freddie began making involuntary noises that resembled a dog or cat, which caused him great difficulty at school. When agitated or excited, Freddie imitated others' words and actions. Although his symptoms briefly lessened in intensity, they always returned with renewed force and additional involuntary movements and behaviors.[20]

Mahler and Rangell observed Freddie for more than two and a half years. Although they wrote that it was likely that Freddie evidenced a "constitutional inferiority of the subcortical structures producing physiological dysfunction," they insisted that this in itself was insufficient to account for his symptoms. Rather, the organic component "renders the individual defenseless against overwhelming emotional and psychodynamic forces." Freddie's motor and vocal tics were a "psychiatrically determined . . . pathological attempt to solve a conflict situation."[21]

Freddie's medical history revealed a number of striking and puzzling organic features. Mahler and Rangell suspected, but could not verify, an earlier encephalitis, which given Mahler's view should have excluded her diagnosis of tic syndrome. Tonsillitis, resulting in a tonsillectomy at age four, and frequent ear infections suggested a history of streptococcal infections. Most interesting was that, when examined for a month in 1941 after an episode of severe tics, Freddie's temperature daily rose to between 100°–101° F without any reason. This was accompanied by an increase in his pulse rate to 110. The temperature generally elevated at mid-day, and once it reached 103°. Freddie sweated excessively. He was obese, which his examiners attributed to his insatiable appetite. Freddie's Rorschach test indicated that he suffered from a mild neurophysiological disorder and that his tics might be due to an epileptic condition.[22]

Unable to explain Freddie's peculiar somatic symptoms, Mahler and Rangell focused their analysis and treatment on the psychological. They attributed Freddie's obesity to his mother's having overindulged him with food, "compensating for her conscious as well as unconscious guilt feelings" because "Freddie was an unplanned, unwanted child" who the mother had attempted to abort during pregnancy. Freddie's nausea with occasional vomiting was attributed to hysteria.[23]

The onset of Freddie's tics was connected to his ambivalent feelings toward his older brother Gilbert. In seeking to obtain approval from Gilbert, wrote Mahler and Rangell, Freddie imitated the eye blinks of Gilbert's close friend and play companion. This, the therapists explained, satisfied an unconscious wish that Gilbert would view his younger sibling as a substitute for the blinking companion. When Gilbert ignored his younger brother in favor of others, Freddie's involuntary movements exacerbated. Freddie was also angry at Gilbert for abandoning him and developed night restlessness, which, according to his psychiatrists, was an attempt to seek revenge against Gilbert, with whom he shared a room. Simultaneously, this provoked guilt in Freddie, which only intensified his involuntary movements. "Gilbert can't study because of my movements," Freddie admitted, "and might even fail in school."[24]

Freddie displayed no coprolalia. Yet this very lack of obscenity— Freddie "kept his own vocabulary spotlessly free of any dubious words and shunned and avoided such expressions by others"— convinced Mahler and Rangell that he was "defending himself against the latent existence of 'mental coprolalia.'" Finally, according to his therapists, Freddie was addicted to excitement. He went to the movies as often as possible and "devoured" comic books and action stories. It was in these instances that Freddie's imitative tendencies (echolalia and echopraxia) manifested themselves most persistently.[25]

Having identified the centrality of psychological mechanisms to Freddie's behavior, Mahler and Rangell, nevertheless, were not sanguine about the long-term impact of psychotherapy. Reporting that Freddie's symptoms had improved during psychotherapy, the two therapists did "not wish to convey the impression that they believe the course of the disease has been checked, or that any permanent or basic changes have taken place." Maladie des tics was a relentless disease and Freddie's prognosis, they concluded, was "unfavorable," because of "a substratum of organic disease or deficiency, the dominance of and lack of control over the subcortical system of expressional motility, and the inability to retain inner stimuli without discharge."[26]

On the one hand, Mahler and Rangell insisted that organic factors merely provided the soil in which childhood conflict might manifest itself in vocal and motor tics. On the other hand, the failure of psychoanalytic therapies to eradicate these symptoms was laid at the feet of their organic cause, which Mahler and Rangell argued was insufficient, in the absence of psychic conflicts, to bring on the condition in the first place. As contradictory as this reasoning may seem in retrospect, it appears to have not lessened the enthusiasm for psychoanalytic interventions in cases of convulsive tics. For whatever limitations psychoanalytic therapeutics displayed in terms of curing these conditions, it was, at least, a fully elucidated system that could be called upon to explain each symptom displayed by those diagnosed with convulsive tics. With its self-proclaimed psychosomatism, psychoanalytic interventions appeared to be integrating whatever organic elements contributed to tics. From that perspective, psychoanalysts were able to lay claim to both the psychic and somatic traditions.

Although Mahler continued to allow for an organic role in tics, these concessions seemed to play an increasingly conventional function, which permitted her to move on to the centerpiece of her concern, the psychogenic features of tic. In 1944, with Rangell now on active military service, Mahler published a second article on convulsive tics.[27] Selecting the case of her patient "Teddy," Mahler explained how early childhood motor movements and vocalizations were transformed from impulsive to compulsive tics. Teddy reported that at first "I twinkled because I saw it in the movies and then because Johnnie did it, the friend of my big brother and later I couldn't help blinking any more." Teddy, wrote Mahler, succeeded replacing his eye blinking by an arm tic and was proud of his achievement. But at this point, Teddy, now nine, developed animal-like grunting, barking, and squealing noises accompanied by echolalia and echopraxia. These, she explained, were the precursors of idea and cursing tics. Vocal tics, imitative phenomena, and impulsive cursing were psychoanalytically no different from gestural tics, such as blinking, head movements, and shoulder shrugs. Coprolalia (cursing) was merely the final logical development of the earlier motor and vocal tics. It represented a condensation of the involuntary repetitive vocalizations and animal sounds often made by small children. These were generally followed, even in normal four- or five-year-old children, by erotic and provocative "bathroom talk." As children mature and attempt to restrain their motor movements, "these eroticized verbalizations turn into . . . compulsive and repetitive motor symptom[s]," which manifest themselves in blurting out curses. Therefore, early bathroom humor was actually an intermediate stage ending in coprolalic tics. Echo behaviors were "compromise solutions" of a ticcer's imitative tendency and of an instinctual repetition compulsion, which became exaggerated when expression was thwarted. But coprolalia eventually emerged because it allowed the simultaneous release of "oral, anal, and phallic, libidinal and aggressive tendencies" that "appear in equal proportion in the usual four letter words of coprolalic tics."[28]

In July 1945, Mahler edited a collection of articles she had solicited for the journal *The Nervous Child,* which she conceived as serving as an overview and for psychiatric diagnosis and treatment of pediatric tic syndrome.[29] The contributors to this collection, *Tics in Children,* included psychiatrists, neurologists, psychologists, social workers, and an art therapist, most of whom worked with Mahler at Manhattan's New York State Psychiatric Institute and Hospital. Authors explored childhood tics from their different professional perspectives, but within a psychoanalytic framework. *Tics in Children* was divided into three parts—recent literature on tics, pathology, and therapy—and explored topics including epidemiology, differential diagnoses, reaction to restraint, neurodynamics, physiology, Rorschach records, and parent-child relations. The collection elaborated and endorsed Mahler's psychosomatic theory of tics, with most of the contributors drawing directly on Mahler's patients, particularly her cases of Freddie and Teddy. To these "typical cases" Mahler and her colleague Irma Gross contributed a third case, that of Pete, to the collection.[30]

Mahler had shared with all of the contributors her forthcoming follow-up study of ticcing children written with her interns, Jean

A. Luke and Wilburta Daltroff.[31] It is important to point out that the paper was required reading for all the contributors to *Tics in Children* because it had laid out Mahler's definition of the psychosomatic "tic syndrome." Placed in this context, *Tics in Children* was an elaboration rather than investigation of Mahler's psychosomatic views. In her introductory remarks, Mahler defined terms and presented her overall conclusions. Why was it, Mahler asked, that only some hyperactive children developed tics, whereas the majority did not? Why did transitory tics, so common in small children, develop into involuntary motor tics in some? The answer, Mahler insisted, was to be found in her research. It revealed that tic syndrome was a combination of constitutional factors set off only in those children who simultaneously experienced too much or too little environmental interference. As her case histories indicated, the guilty parents, more often than not, were mothers who interfered with their children's motor expression, inhibiting repetitious impulses, aimless motor habits, and excessive emotional demonstrativeness. As a result of these parental restraints or, alternatively, lack of restraints, constitutionally predisposed children later were unable to assert normal control over their motor activities.[32]

Each of the chapters that followed underlined Mahler's thesis. The opening essay, an overview of recent literature on tic by psychoanalyst Samuel Ritvo, explained that environmental and family factors almost universally were at the root of tic syndrome.[33] Ritvo cited Mahler, Luke, and Daltroff's forthcoming study as reporting that 88 percent of cases of childhood tic syndrome resulted from parents' overindulgence, infantilization, and overprotection.[34] As an example Ritvo pointed to Mahler and Rangell's 1943 case of Freddie, whose mother had overwhelmed him with her anxieties about his health because of her guilt for having attempted to induce an abortion when pregnant with him.[35]

Bernard L. Pacella, who had administered and interpreted Freddie's electroencephalogram,[36] found little evidence for a physiological basis for tics. Like Mahler, Pacella concluded that those tics not caused by an identifiable infection were psychogenic.[37] In the next essay, Cleveland psychologist Thesi Bergmann drew on Mahler and her colleagues to suggest that temper tantrums served as a stage leading to "the crystallization of tics" in many children.[38] Psychiatrist Paul H. Hoch argued that ticcing children suffered from developmental immaturity because their rhythmic, habitual, and repetitious emotional discharges resembled those of infants.[39] Thus Hoch seemed to suggest that children who had similar psychological conflicts, but lacked this physiological underdevelopment, would manifest different symptoms, such as epilepsy or bowel disorders, from those displayed by ticcers.[40]

Following Hoch, psychotherapist Esther Menaker selected a seven-year-old patient, Alice, as an exemplar of Mahler's claim that parental indulgence exacerbated and enabled tics. "In the case of Alice," wrote Menaker, "there was not only an overtolerance for her hyperactivity in the prekindergarten and kindergarten years, but even an admiration of it, which precluded the possibility of its being understood by her parents or teachers as the expression of neurotic conflict." Only later, when social conditions required self-restraint, Menaker explained, was it evident that Alice's body movements and animal-like sounds were caused by unresolved emotional problems. Although she behaved like a "perfect lady" in Menaker's office, if she was forced to wait for even a short period of time, Alice reacted by shouting loud animal noises and jumping on the office furniture. Alice could not tolerate waiting because, in her therapist's words, "waiting meant that I was devoting my time to someone else." This situation was extended to every aspect of Alice's social relations, so that Menaker concluded that Alice's impulsive motor activity, barkings, and gruntings revealed a deep jealousy.[41]

Jealousy of others, however, was merely the manifest content of Alice's subterranean penis envy—her wish to be a boy. "Interestingly enough," related Menaker, "one of her first drawings was of a house which was tall and tower-like in structure, had no doors or windows, but only a chimney on top with some smoke coming out." As Menaker interpreted it, Alice's drawing "seemed quite obviously a penis symbol," representing her attempt to express and symbolize her unrealizable and unconscious wishes that previously had been acted out in her movements and vocalizations. This interpretation was confirmed by a rereading of Alice's symptoms: "We must stress," Menaker wrote, "the fact that much of Alice's behavior was of a decidedly erotic character, not only in the sense that the bodily movements as such were erogenized, but that she literally masturbated while tossing or throwing her body around and while performing acrobatic feats." Moreover, Alice "liked to sit on people's laps, fondle collars or beads, rub her hands along one's legs or stroke one's hair." This behavior confirmed what Alice's drawings illustrated; she was "an oversexed child."[42] Her sex role confusion was evident in Alice's "actively seductive behavior" because she characteristically rejected feminine passivity for an active masculine role.[43] Apparently, none of these insights did much to relieve Alice's symptoms, which Menaker believed would only abate if Alice would successfully identify with her mother.[44]

The next essay, by psychologist Zygmunt Piotrowski, who had evaluated Freddie, explained how the Rorschach test could reveal the psychogenic issues that underlay childhood tics.[45] Ruth Henning Latimer, a social worker, contributed the final essay in the section on pathology. Focusing on the role of parent-child relationships in children with tics, Latimer claimed that parents of ticcers were "inflexible, rigid, and anxious." In most cases, Latimer insisted, it was "mothers particularly," who "restricted the patients' freedom in play and activities."[46]

The final section of *Tics in Children*,[47] entitled "Therapy," contained Mahler and Gross's extended analysis of an eleven-and-a-half-year-old named Pete, whom the authors had selected as a patient with a typical example of tic syndrome.[48] Pete's case, along with Mahler and Rangell's 1943 case study of Freddie, would become emblematic touchstones for psychoanalysts as they wrestled with the diagnosis and treatment of tics for the next quarter century.

Pete's initial symptoms—throat clearing, rolling his eyes, mouth twitching, yawning, and stretching his neck—appeared when he was three or four. These continued, waxing and waning, even disappearing entirely for a year, until Pete was ten. Then the symptoms reappeared with increased intensity and with frequent vocalizations. This combination of tics and uncontrolled vocalizations expressed themselves on Pete's entire body, including his face, diaphragm, trunk, and extremities, and incapacitated him. Pete was admitted to Mahler's care at the Psychiatric Institute and Hospital in January 1945.[49]

Mahler and Gross's psychoanalytic examination convinced them that Pete's behaviors resulted from his anxious grandmother and his mother having "overindulged and infantalized the boy." This served as evidence for Mahler's main thesis that conflict over Pete's motor freedom by his caregivers was typical of children with tic syndrome. Mahler and Gross uncovered three other typical factors in Pete's case—phobic fears, frequent occurrence and discussion about accidents, and over-attention to his body.[50]

In their treatment, which included free association and dream interpretation, initially Pete was allowed to introduce any subject for discussion, but eventually Mahler moved the discussion to what she believed to be Pete's "basic problems," which were sexual in nature. Mahler and Gross interpreted Pete's initial nonchalance and proclaimed disinterest about sexual issues as a signal of his repressed concerns with sexuality:

> In discussing sexual material Pete adopted an attitude of objective matter-of-fact interest and inquisitiveness. For instance, he asked a series of questions of a scientific nature, starting with the structure of snow flakes, which were then falling, going over the crystallography in general, and ending with his statement: "The more you know of this stuff, the more interesting it is. It is not like sex. My cousin used to tease me because I didn't know about sex . . . so now I know it, and there is nothing much to it. It's not so interesting." In fact, we may assume that Pete really meant that sex *is* the subject which seems inexhaustibly interesting to him.

As with the noncursing Freddie, Mahler uncovered repressed coprolalia in Pete. His sexual fantasies and preoccupations, according to Mahler and Gross, were both forerunners of and the "noncrystallized equivalents" of involuntary cursing.[51]

Pete grew depressed and discouraged because his tics did not abate, and complained to his mother that psychiatry was "witchcraft and superstition." When questioned by his therapist about these attitudes, Pete responded, "It's silly that's all. All this talk. How can it help me? It's like the Middle Ages. Those faith healers. That's just like this business." Mahler attributed Pete's outburst to the therapy's success in exposing the unconscious sexual issues that had enabled Pete's tics and the boy's fear of his own sexuality. To prove the point, Mahler asked Pete if he knew how condoms worked. "Because they have no holes, they keep the germs from getting in or out," Pete replied. Pete's use of the term "germs," which the therapists had introduced in the first place, revealed their patient's fears and preoccupations. Their evidence for this was Pete's question about what type of germs caused elephantiasis and dropsy, diseases he had read about in *Life* magazine. His query was interpreted by Mahler and Gross as reflecting Pete's obsession with masturbation: "Pete worried about his swellings, and the probability of his awareness that during mumps [which he recently had] the testes may become involved was interpreted as the motivation for his present preoccupation with such 'swellings' as elephantiasis and dropsy. He admitted knowing that his testes could have been affected while denying any concern with his characteristic defense: 'I didn't have to watch them; I figured that if they got it they'd swell up and hurt just the way my face did.'" Pete's refusal to acknowledge his fear that mumps could cause his testes to swell protected him from discussing other swellings of his genitals—erections. When Mahler connected these issues, Pete "characteristically felt like dropping the disquieting subject, and went on in the next interviews with minute reports on his tics."[52] Mahler and Gross saw this reaction as proof of Pete's fear of his own erections.

During his free associations Pete had connected putting his hands on dangerous places with having discovered condoms in his parents' dresser drawers.[53] Pete's therapists believed that this association again pointed to the issue of masturbation. First, they pointed to Pete's conspicuous preoccupation with his body, moving from concerns about swelling to his hands. Second, Pete's report of his mother's statement that it was "too late" for her to get pregnant revealed his own displaced anxieties about using up his reproductive energy through masturbation. Third, Pete tended to turn all discussions toward sexual matters. Next, Pete was ambivalent toward adult authority, which restrained his freedom while protecting him from unwanted sexual impulses. Finally, Pete's dream of his arm moving back to touch a warm radiator was, for Mahler and Gross, clearly symbolic of masturbatory ambivalence.[54]

Succeeding interviews more explicitly elicited questions from Pete about erections and, finally, after much prodding, masturbation itself. Pete was also informed that his conflicted feelings about masturbation could have led to his tics. He remained skeptical: "You told me the tics came because of the mixed-up feelings and lots of those feelings have to do with the sex business. Well," Pete reminded his doctors, "I've had tics since I was about 3 years old. How does that fit in?"

At this point, reported Mahler and Gross, Pete's tics had abated with only residual eye deviations and eyebrow elevations. Several weeks later, Pete's facial tics returned. His therapists attributed this exacerbation to their treatment of a new patient, resulting in a sibling rivalry as Pete and the new patient competed for psychiatric attention. In the end, by the summer of 1945, Pete was less satisfied than his caretakers, who pronounced him cured: "Yeah. The little movements are back again. But, the big ones that annoyed me, are still all gone. If those could be cured, why shouldn't these?"[55]

Follow-ups Define Therapeutic Goals

In October 1945 Mahler, Luke, and Daltroff published the results of a follow-up study of the eighteen children who had been treated for tic at the Children's Ward of the New York Psychiatric Institute and Hospital.[56] In order to "find scientifically comparable results in this complex problem," the trio retrospectively limited their sample to those children who fit their latest definition of tic syndrome.

Mahler and her colleagues reported a number of epidemiological findings from which they constructed clinical conclusions. First, they noted that tic syndrome was much more often found in boys than girls. Of the eighteen follow-ups, twelve were Jewish, leading Mahler to conclude that "whatever predisposing part the constitutional and racial factors may play in motor neurosis, they fit into a rather specific national and cultural family pattern which our study revealed was frequently found in the tic children and seemed to facilitate development of psychomotor neurosis." They also pointed out that many children with tic syndrome had a symbolic position in their families that had exacerbated their tendency to tic. Twelve were only children, seven were "the baby" in the family, ten were the first surviving child, but only three were middle children and only one was the younger of two children. From this and other data, including the high percentage of breast-fed and overweight ticcers, Mahler and her colleagues concluded that many of these children were overindulged in the oral stage and continued to depend on their mother's feeding until six or even eight years of age. Further evidence in support of this overindulgence was that 80 percent of their clinical cases displayed temper tantrums, and in 50 percent of their follow-up cases mothers cited temper tantrums as one of the prominent reasons for bringing their children back to the clinic.[57]

The lesson to be learned from all of this data was that whatever predisposing organic factors existed, the tic syndrome appeared only when caregivers, generally mothers, failed to set the bounds of activity early in the child's life. Such children, as they entered the psychosexual minefield of puberty, were at high risk for tic syndrome. But what did all this teach a clinician faced with a child with putative tic syndrome?

Because of ticcers' early fixations on psychosexual development and their problems with "ego synthesis," the authors focused on three therapeutic strategies. First, psychotherapy should aim at making unconscious mechanisms conscious, to make the child aware of the issues that led to the tics. Second, the pathological familial environment should be eliminated wherever possible. Finally, these children should be given an opportunity to channel their motor urges in positive ways, such as through exercise. Beyond these, Mahler and her colleagues noted that in severe tic cases psychotherapy should be followed by temporarily placing the child "in a good country school, or a country foster home," so that the child could be separated from the family environment, which provided the soil for the tics to develop in the first place.[58]

Mahler, Luke, and Daltroff admitted that their follow-up data gave the impression that the prognosis for tic syndrome depended on numerous factors that they had not discussed, but they promised this in a later paper.[59] That study, coauthored by Mahler and Luke, appeared in *The Journal of Nervous and Mental Diseases* in May 1946. While the earlier follow-up study had examined the charts of eighteen ticcing children, Mahler and Luke reduced the number to ten, excluding all girls from the sample. Although Freddie, Teddy, and Alice had been part of the earlier follow-up study, they were dropped from this study. Selecting their cases for follow-up study, Mahler and Luke explained that they had discarded many cases of proven organic involvement or psychosis in which tic syndrome was only a complicating factor. Thus Freddie, Teddy, and Alice, whose cases already were and would continue to serve as exemplars of Mahler's tic syndrome, were eliminated from her most detailed examination of the outcome of psychotherapy. She had decided that, strictly speaking, none fit her criteria for tic syndrome in the first place. Pete, it appears, was not considered in either of the two follow-up studies.[60]

Mahler and Luke's most sobering finding was that psychotherapy often failed to remove or even ameliorate a patient's tics. According to their study, there was "no direct correlation between recovery from the tic syndrome and the length and method of psychother-apy." Moreover, they had uncovered no connection between "the thoroughness of treatment and good therapeutic results." How-ever, Mahler and Luke claimed that this lack of a definite correla-tion between recovery from the tic syndrome and psychotherapy was not surprising. Rather, they were convinced that such a result was "indeed to be expected if our concept of the tic syndrome as an organ-neurotic disorder of the neuromuscular system is cor-rect." This was because psychosomatic tics impaired the ego to such an extent that there was a "veritable incontinence of emotion" since "erotic and aggressive instinctual impulses" continually es-caped through the tics and vocalizations, which themselves exac-erbated the illness. The condition became chronic as these instincts were "dammed up," causing overstimulation of the already sus-ceptible parts of the body. Because child ticcers were fixated on psychosexual developmental issues, a vicious circle was established. Here Mahler and Luke reminded their readers that Freud too had concluded that a "vicious circle of organ-neurotic symptoms" was "not directly accessible to psychoanalysis." Thus, what might ap-pear to the uninitiated as a failure of psychoanalytic method actu-ally had dialectically confirmed its most profound theoretical in-sights.[61]

Mahler and Luke, however, did not conclude that their follow-up results indicated that psychoanalysis should be jettisoned in cases of tic. To the contrary, psychotherapy retained its central role, once that role was clearly understood. They explained that the disap-pearance of ticcing symptoms was not the chief aim of therapy. That was because the follow-up data revealed that "disappearance of the tics was not always accompanied by improvements in the total personality." In fact, in some cases, overall adjustment was reached even though a few tics persisted, whereas in other cases where the tics disappeared, the patient remained psychologically disabled. This confirmed the wisdom of the psychoanalytic concept that the "symptoms serve the purpose of discharging dammed-up instinctual impulses in a pathologic way. Thus, the tics represent a kind of morbid release, a safety valve for release of tension."[62]

From this perspective Mahler had reconstructed the apparent failure of her method to relieve vocal and motor tics as a success, whereby the goal of treatment was no longer merely symptomatic relief, but avoidance of a greater psychopathology that might follow if the patient's need for release from dammed-up libido were restrained by removing the ticcing behaviors. "The aim of therapy," concluded Mahler and Luke in 1946, "should be to give the child's ego an opportunity for gaining perspective concerning those impulses, and the necessary strength to cope with them." Even though the tics might never be removed, Mahler was encouraged. Nevertheless, she insisted that if a favorable outcome was to be achieved, psychoanalytic treatment of tic syndrome must be instituted before habits became imbedded in the child's unconscious. Early intervention was essential; doing nothing was dangerous because, Mahler warned, even when tic syndrome seemed to disappear, it was usually replaced by a severe personality disorder.[63]

In 1949 Mahler wrote about tic syndrome for the last time. Reviewing sixty cases of children diagnosed with tic, Mahler reaffirmed her belief that tic syndrome resulted when overprotective mothers infantilized their (overwhelmingly male) children who already had a constitutional tendency toward hypermotility. Mahler sought to refine her earlier findings, confirming again in this study that psychotherapy appeared to have little impact on ticcing symptoms. Attempting to separate the therapeutic successes from the failures, Mahler, who initially had segregated infectious causes of tics to establish a psychogenic type she called tic syndrome, drew a further distinction between "tic disease" and "impulsive tiqueurs." Tic disease, according to Mahler, was a psychosomatic "organ neurosis of the neuromuscular apparatus" in which the tics "in themselves represent the central and essential disturbance." Impulsive ticcers, by contrast, were those whose tics were a sign or symptom of other psychopathologies. Although the grounds for such a distinction among her patients were far from overwhelming, Mahler erected the categories to help explain, *ex post facto*, why some ticcers got better and others did not. Surprisingly, she claimed that the prognosis was "relatively favorable" for those with tic disease in contrast with those for whom tics were merely a symptom of a deeper disorder. In the end, Mahler, like Gilles de la Tourette and Charcot, seemed unable to explain either the recovery from tics or their persistence. Like her French predecessors, Mahler constructed a self-evident solution—those who were cured suffered from a different disorder from those who were not.[64]

Although the aim of Mahler's typological distinction was to alert therapists about the types of children who might experience the best clinical outcomes, Mahler's discussion of her patients betrayed extreme frustration. In fact, if one did not know better, one might read her language as an indictment of the afflicted. Mahler declared that those children with tic disease were the most rigid and resistive to giving up defensive mechanisms. They "lack spontaneity and initiative" and "defend themselves by overcompliance." By doing so, children with tic disease "have complied with their mother's wish for their remaining vegetative creatures, with no will and intention of their own." By contrast, the "impulsive tiqueur, . . . like a delinquent, is artful in evading therapeutic interference by 'acting out' and projection mechanisms."[65] It seemed that Mahler had concluded that the greatest obstacles to effective treatment of tics were the patients and their mothers, rather than the disorder itself.

In the late 1970s and early 1980s, Margaret Mahler, now the doyenne of pediatric psychiatry, looked back at her work on ticcing children in the 1940s.[66] In these recollections Mahler emphasized that her most important finding had been that some ticcers, "whether or not they retained their tic symptoms in later life, became psychotic at puberty." This assertion, though couched in psychoanalytic metaphor, mirrored Gilles de la Tourette's claim that maladie des tics was a progressive disease that led to debilitation and often to insanity. "My thinking on this issue," explained Mahler, "led me to understand the steady bombardment of the ego by the tic paroxysms in terms of a weakening of the ego. The ego was simply unable to master these tics, which therefore achieved involuntary discharge through motility." This debilitated ego, Mahler concluded, "proved unable to cope satisfactorily with the new and stressful tasks brought on by puberty."[67] Gone was any acknowledgment that organic conditions underlay the production of tics, even though by this time haloperidol was being used with success on ticcing patients.

In retrospect, psychoanalysts themselves exposed so many anomalies in their explanations and treatments of tics that, from today's perspective, it is difficult to understand their continued rationalization of each contradiction as if it were an affirmation of a theoretical claim. "How does one explain," asked psychiatrist

Arthur K. Shapiro in 1978, "the startling and somewhat dramatic diversity of observations and theories among psychoanalysts?" Answering his own question, Shapiro offered several possible answers, including the possibility of ascertainment bias, "that the described psychopathology might have characterized a particular patient but was unrelated to Tourette syndrome. The logical fallacy was to generalize the results from a single patient to all patients." Beyond this, Shapiro noted that psychoanalytic "theories and observations might be prone to theoretical faddism. Once a body of theory becomes popular, there is a tendency for subsequent clinicians to find and conform to them." But "another possibility" was "the use of projection, in which disowned inner fantasy is attributed to or projected onto the patient as the explanation for clinical phenomena."[68]

Although Shapiro did not provide any evidence for his psychoanalyzing the analysts, he could have later found confirmation in Margaret Mahler's memoirs. "I came far too early—nine months and six days after the wedding," Mahler wrote, "and was very much unwanted by my mother, who was a mere girl of nineteen at the time . . . In her anger" my mother "had as little to do with me as she could." Like so many of the sibling conflicts that Mahler "uncovered" in her ticcing patients, she remembered that "the arrival of my sister, Suzanne, four years after my birth only aggravated my sense of maternal rejection. She was very much a 'wanted' child, and she awakened our mother's maternal instincts—instincts that for me, had lain dormant."[69]

For all the currency that Mahler and her colleagues had placed on the interpretation of language, in the end, Mahler and the other psychoanalysts who relied on her claims proved unable to listen to what their pediatric patients attempted to tell them. When placed in the context of Mahler's earlier (1946) admission that her retrospective study had shown that there was no direct correlation between recovery from the tic syndrome and the length, method, or thoroughness of treatment, her pediatric patient Pete's doubts about the sexual emphasis in his treatment seem less like resistance than reason.

Haloperidol and the Persistence
of the Psychogenic Frame

8

Although Margaret Mahler abandoned the study of tics in the late 1940s, those psychoanalysts who came after her relied on her published cases and conclusions, holding them up as models to be emulated and texts to be revered. For the next two decades psychoanalysts would repeat, as if it were a medical mantra, that tic disorders resisted psychoanalytic interventions because "the role of the tic" was "the last 'desperate defense against psychosis.'"[1] As with Mahler, such admissions heightened rather than lessened claims that psychoanalytic theory exposed the underlying forces that produced these involuntary movements. Certainly, it reinforced the belief that only those trained in psychoanalysis possessed the skills essential for helping ticcing patients negotiate the psychological shoals that threatened to sink them into a psychosis. Adherence to this belief was so pervasive that the discovery in the 1960s that the antipsychotic drug haloperidol was an effective tic suppressor did not challenge most psychiatrists' assumption that psychosexual conflict was the underlying cause of motor and vocal tics.

Even before Mahler had penned her final thoughts on tic syndrome, her influence was evident in the views of the contemporary North American and British psychoanalytic community.[2] In 1948 Eduard

Ascher, professor of psychiatry at Johns Hopkins University, reviewed five cases of "maladie des tics" and concluded that motor movements, echolalia, and coprolalia were "related to certain attitudes the patient had toward one or both parents." At the same time, the tic served as an attempt to suppress the expression of these attitudes. Like Mahler, Ascher saw tics as symptoms of more profound psychiatric disturbances, including schizophrenia. Ascher tied the failure of psychotherapy to alleviate tics to "the presence of some central nervous system pathology," even though he offered no evidence to support his assertion.[3]

By the late 1950s, Manhattan psychoanalyst Z. Alexander Aarons, drawing almost entirely on Mahler, described a fifteen-year-old patient's twitching and yelping as "overdetermined symptoms." When the boy was incapable of suppressing his twitches, he despaired about the damage that he might do to his body. "The crucial point," wrote Aarons, was that the patient was convinced that "he had 'ruined' himself by his twitching. He spoke of himself as a 'twitching wreck.'" Aarons was certain that the boy's "unconscious thought was that he had *ruined* himself as a result of his masturbation." The boy's "tension and 'urge' attendant upon his twitching and yelping force us," wrote Aarons, "to the conclusion that this patient had highly eroticized his whole voluntary neuromuscular system and that the twitching was a masturbatory equivalent." Twitching, therefore, was a substitute for masturbation. It also became a symbolic mechanism for the patient's fear of "submission to his passive feminine wishes." These feminine wishes were evident, among other things, in the boy's repeated demand for a neurological examination with the hope, which Aarons portrayed as another fantasy, "that something physical might be found and excised." Related to this was the boy's fear that his symptoms would result in incarceration in an insane asylum. This fear, according to Aarons, was "the equivalent of his fears of being someone else and of losing control as the result of masturbation." However, it was difficult for the boy to understand the connection between masturbation and his fears.[4] Like Mahler's pediatric patients, complaints about ticcing symptoms were viewed by the analyst as additional evidence of unconscious sexual conflicts. Unable to rid their patients of tics, psychoanalysts devalued the tics themselves as a possible source of distress.[5]

Organic Interventions

Although psychoanalysis would continue to dominate the treatment of tics throughout the 1950s and 1960s,[6] a number of practitioners, citing their success with a variety of behavioral, surgical, and pharmacological interventions, were persuaded that motor and vocal tics were variously enabled by developmental, infectious, and organic factors. Nevertheless, either out of deference to the cultural power of psychoanalysis or due to an inability to see beyond it, these findings often were framed in psychoanalytic language. As a result, their implications were often muted and largely ignored.

Drawing on the popular postwar view that stuttering resulted from forcing left-handers to switch to using their right hands, a number of studies found a parallel in the etiology of tics. A 1948 French report claimed that most ticcers were left-handed, even though their left-handedness was not always evident because almost all left-handers had been forced to write with their right hand. Based on observations of two left-handed ticcers, the authors were convinced that it was possible to ameliorate tics or even make them disappear simply by returning the patient's left hand to its naturally dominant role.[7] These claims were supported by a 1954 psychiatric report, also from France, claiming that the best way to inhibit the procession of tics was through education that included freedom of choice to use the left hand.[8]

Given their psychoanalytic assumptions, psychiatrists often neglected to explore the pharmacologic implications of chemical agents that reduced tics. In 1950 French psychiatrist J. N. Lanter reported a rapid cure of a thirty-nine-year-old male patient's multiple tics, coprolalia, and echolalia with the drug parpanit, which today is sometimes used as an antidepressant. Lanter was convinced that it was not the drug's action but the patient's psychological fantasy about the drug's symbolic role in treatment that resulted in his improvement. The tranquilizing action of the parpanit, wrote Lanter, suppressed the man's aggressive feelings toward his father.[9]

By the 1950s most French psychiatrists united behind a psychoanalytic view. Swiss and German practitioners, however, building on the earlier work of Raymond de Saussure and Erwin Straus, continued to see motor and vocal tics as having an infectious etiology.[10] Based on analysis of two cases, Swiss medical professor G. de Morsier argued in 1951 that the maladie des tics de Gilles de la Tourette was most definitely an organic/postinfectious disorder. The first case was a forty-year-old male butcher with uncontrollable motor and vocal tics, including, at ten-second intervals, extremely loud barking sounds.[11] Morsier tied these symptoms to an earlier infection following an episode of bronchial pneumonia. Each waxing and waning, reported Morsier, seemed to be tied to flare-ups in infection. Morsier was certain that the cause of the man's symptoms "seems likely to be the result of an inflammatory attack of the brain, but particular aspects of the crisis must be considered to originate in the temporal lobe." The second case was a sixteen-year-old girl whose initial symptoms began in 1944 after her father was shot by the Germans. But much more robust symptoms followed her illness with tonsillitis at age thirteen. After that, the girl developed tics and bizarre compulsive behaviors including "having to touch the ground with her hand each time she took a step." Although Morsier did not claim that tonsillitis had caused the symptoms, he was convinced that it had aggravated them.[12]

Morsier believed that, when combined with other recent reports, these two cases demonstrated that Gilles de la Tourette syndrome extended from predominantly respiratory and anatomical neurological crises, most likely resulting from encephalitis. Support for this conclusion, according to Morsier, was found in experiments demonstrating that stimulation of the hypothalamus caused tics, while stimulation of the temporal lobe caused vocalizations. Morsier emphatically rejected the psychoanalysts' claims that tics resulted from "regression": "Like concepts of degeneration of earlier times, the 'dissolution of function' of today, in trying to explain everything, explains nothing at all."[13]

Like Morsier, those who argued that multiple tics and coprolalia had an organic etiology increasingly centered their discussions on damage (whether by infection sequels or by other means) to the structures of the basal ganglia, particularly the caudate nucleus and the putamen, which form part of the striate cortex. For instance, in 1952 a French neurologist, explaining that different involuntary movements could be produced experimentally by stimulation of specific striatal cells, argued that tics, choreas, and other involuntary movements resulted from specific but different striatal lesions. The location of a lesion thus determined the particular manifestation of an involuntary movement.[14] Other studies investigating the effects of pharmacological interventions supported this claim.[15] From either perspective, evidence was increasing that, in the words of a Genoese neurologist, "multiple tics and Gilles de la Tourette's disease . . . certainly compromised the structures of the basal ganglia."[16]

The strongest scientific confirmation of the role of striatal structures in ticcing behaviors came from two German neurologists, Johan Ludwig Clauss and Karl Balthasar, who reported in a 1954 paper that an autopsy and histological examination of a patient diagnosed with Gilles de la Tourette's disease had exposed lesions in the striate cortex. Histology (study of cells) had revealed a great number of intact but smaller than normal cells. This pattern differed from that of chorea, where the cells conserved their size, but were massively reduced in number. The authors theorized that a combination of constitutional underdevelopment and lesions of the striate cortex, which irritated the routes of sensory conduction of the spinal cord, had produced the tics. Along with typical motor tics there were various compulsive symptoms such as urges to move, imitative behaviors, and involuntary cursing. Contrary to

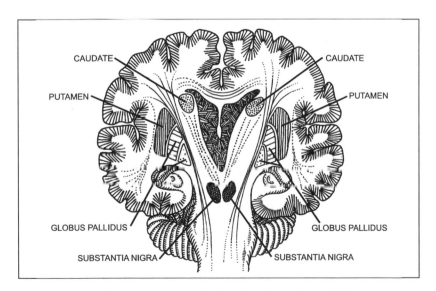

Figure 8.1 The brain structures called the basal ganglia, which regulate motor movements. Disruption of the basal ganglia circuitry has been associated with the involuntary movements displayed in Tourette syndrome.

Straus, who believed that these movements resulted from chronic Sydenham's chorea, Clauss and Balthasar were persuaded that the abnormalities they had uncovered pointed to a physiological underdevelopment as the key to these behaviors.[17]

In a second paper, published in 1957, Balthasar cited the autopsy of a forty-two-year-old male patient whose brain resembled that "of a one-year-old child." It was a well-known neurological principle, wrote Balthasar, that an immature striatum resulted in reduced ability to control motor movements. The patient's childlike motor movements resulted from the underdevelopment of the striatum, which was, according to Balthasar, "the anatomical foundation" of the hyperkinetic movements found in Gilles de la Tourette's disease. Balthasar was uncertain whether this cellular underdevelopment was the result of a congenital condition or whether it resulted from an infection such as encephalitis.[18] In either case, he was certain of the physiological cause of the symptoms.

For a variety of reasons, not the least of which was the German language in which these articles appeared and the continued resilience of psychoanalysis, none of these continental studies made a noticeable impact on Anglo–North American practitioners' understanding of the etiology or treatment of multiple motor tics and involuntary vocalizations in the 1950s. Moreover, given its long history, the claims that some tics were sequels to infection were not seen as a particular threat to the psychoanalytic paradigm because, since the 1920s, psychoanalysts simply segregated these cases from the mainstream as either atypical or not really a case of tic syndrome.[19] Not surprisingly, the belief that involuntary motor movements were a postinfectious reaction was most often endorsed, outside North America and Britain, in places where rheumatic fever remained persistent.[20] In contrast, the movement toward an organic view in North America began with a variety of medical interventions. Some of these, like the treatment of chronic sinusitis, updated earlier claims, and others, like brain surgery and the use of pharmacologic agents, reflected new directions in neurology and psychiatry.

Building on Laurence Selling's 1929 account of curing tics in three patients by removal of their infected sinuses,[21] an Oregon pediatrician, Edward E. Brown, reported that observation of thirty-four patients with habit tics—blinking, sniffing, hacking (throat clearing), strange sounds, mouth jerking, ear pulling, nose twitching, and sucking—convinced him that tics were invariably secondary features of chronic sinusitis, purulent (pus-filled) secretions, and toxemia. Cure or control of tics required treatment of these results of sinusitis. Supplementing his article with photographs of his ticcing pediatric patients, Brown pointed to three kinds of evidence to support his claim that chronic sinusitis was responsible for most tics. The first was that the onset of the tics invariably followed a sinus infection.[22]

Second, Brown's observations convinced him that tics were an initially purposeful reaction to sinusitis. Eye blinks, for instance, were a protective reaction against light sensitivity caused by sinusitis; while nose twitching resulted from a child's switching from one nostril to another in an attempt to sniff up nasal secretions. Both throat clearing and odd sounds like "hem" likewise resulted from efforts to dislodge throat secretions. Ear pulling was the child's reaction to earaches. The purposeless habit spasms or tics developed as a result of these initially purposeful spasms.[23]

Finally, Brown found a correlation between tics and the seasonal variations in the incidence of bacterial sinusitis. Tics seldom first appeared in warm summer months, and established habit spasms tended to ameliorate in the summer. Like chorea and rheumatic fever, which followed a similar seasonal cycle, tics, according to Brown, were connected to streptococcal infections. It was often difficult, wrote Brown, to separate habit tics from chorea; in fact, Brown was convinced that many individuals who had contracted chorea later developed chronic tics. As with rheumatic fever and chorea, Brown was persuaded that in chronic sinusitis, streptococcal toxins penetrated the central nervous system producing twitchings like those found in scarlet fever patients. Looking back over his years of practice and citing his previous publications, Brown found that most ticcers had histories of frequent infections, sinusitis, and previous chorea.[24]

Brown rejected Selling's radical interventions of surgical removal of infected sinuses, reporting "excellent results" with more conservative treatment, including twice daily use of nose drops or spray, the application of hot wet towels and cold cream to the face, and sleeping with one's head elevated. In addition, prolonged reclining, chilling, fatigue, allergens, and emotional upsets were to be avoided. During acute exacerbations of chronic sinusitis, penicillin was prescribed. Using these methods, Brown claimed that in almost all cases the tics improved in a few days. "Several children, removed from school because of merciless teasing, were returned to school in a week or two," including one child whose tics had persisted for years.[25] Although Brown was convinced that neither psychogenic factors nor imitation were important factors in the production of tics, he nevertheless conceded that both of these could exacerbate tics.[26]

Like Brown, those who advocated other types of medical interventions against tics also seemed forced, as if by convention, to acknowledge and include psychoanalytic theory in their discussions. The most important of these was the use of carbon dioxide (CO_2) gas ("carbogen"), in which patients received a series of approximately eight- to twelve-second inhalations of a mixture of 30 percent carbon dioxide and 70 percent oxygen three times per week. The goal was to anesthetize the patient. Each session consisted of about sixty inhalations, and a course of treatment could run from fifteen to two hundred sessions of these inhalations.[27] According to its prime advocate, Hungarian-born American psychiatrist Ladislas Joseph von Meduna, carbon dioxide treatment was based on "the theory . . . that psychoneuroses are disturbances of lower structures of the brain, which disturbances upset or distort the emotional values of the concepts formed in the brain cortex."[28] Treatment was assumed to act on these to strengthen the ability of the "higher" cortical structures to resist the onslaught from below because it "raises the threshold of stimulation of the exercised nerve cell" and "increases the ability of the nerve to conduct trains of impulses." The result was that "increased cerebral inhibition is repeated in the cortex of the brain at every treatment" and that "decreased excitability of the cortex is reproduced by every treatment," allowing the brain "more clearly to balance energy demands with the oxidative process." Meduna reported that carbon dioxide therapy cured 68 percent of a variety of "psychoneuroses," including sexual dysfunction, homosexuality, stuttering, and hysterical blindness.[29]

As with other interventions that might suggest an organic pathogenesis for these symptoms, the reported "success" with carbon dioxide inhalations was interpreted within a psychoanalytic frame and presented as further evidence for the insights of psychoanalytic theory. Returning to Freud for authorization of his claims, Meduna reminded his readers that in the *Outline of Psychoanalysis,* Freud wrote that "the future may teach us how to exercise a direct influence, by means of particular chemical substances, upon amounts of energy and their distribution in the apparatus of the mind. It may be that there are undreamed of possibilities of therapy."[30] Similarly, California psychiatrist J. D. Moriaty claimed that "carbon dioxide therapy facilitates treatment by brief analytical methods through which the experienced therapist may usually uncover rather quickly the infantile trends that generally constitute the core of a neurosis."[31]

Given its compatibility with psychoanalytic approaches, carbon dioxide therapy was easily folded into psychosomatic treatments of ticcing patients. In 1957 British psychoanalyst Richard P. Michael reported that his use of CO_2 inhalations helped overcome a ticcing and cursing patient's resistance to psychotherapy.[32] Building on Michael's account, a number of other psychiatrists adopted CO_2 inhalation therapy to treat tics as an adjunct to other interventions.[33]

The report of Ian J. MacDonald, a British psychiatrist at St. George's Hospital in London, was typical. MacDonald used carbon dioxide inhalations on a nineteen-year-old woman with numerous motor tics, clicking and sucking noises, and "ejaculation" of "phrases such as 'cocky-breasty'" and "cock-a tit." Although her electroencephalogram examination indicated damage to that part of the brain associated with articulation of speech (Broca's area), MacDonald was certain that the underlying issues were hysterical, tied to the patient's mother's unrelenting constraints. MacDonald was uncertain that the woman was approachable through psychoanalysis, however. Michael's (1957) report claiming success in a similar case using CO_2 inhalations persuaded MacDonald to attempt a parallel approach. The twelve treatments, consisting of sedatives (amylobarbitone sodium) and stimulants (methylamphetamine) followed by the CO_2 inhalations, were, as described by MacDonald, completely informed by psychoanalytic theory: "The aim of treatment was to engender a feeling of anxiety due to fear of loss of internal control, and at the same time enable the patient to test her view of myself as an aggressor." The purpose of the sedatives and stimulants was to release the patient's emotions, causing her to experience a "feeling of disruption." This, MacDonald believed, would result in the patient's learning new and more rewarding patterns of behavior, through which she would be able to inhibit her anxiety by MacDonald's "reassuring presence acting in an accepting secure atmosphere."[34]

During the eight weeks of treatment the patient's tics gradually diminished as she simultaneously expressed her disdain for her mother "as the 'cocky-breasty'—a male-female cock of the roost." The patient was discharged from the hospital "free of symptoms, feeling more confidently out-going and less self-conscious." She subsequently took a job as a nursing trainee in a location far away from her mother. Three years later MacDonald received a letter in which the woman reported that she had completed her assistant nurse training, was engaged to be married, and had "not had any of the trouble of which I originally complained and I am very happy with life."[35]

Cutting Ticcing Brains

The 1950s was a time when, especially in North America, surgery was often used to attempt to treat seemingly intractable psychological disturbances.[36] The procedure (lobotomy or leucotomy) was performed on a number of patients diagnosed with Tourette's. The leading advocate of surgical treatment for a variety of psychiatric disturbances was surgeon James Watts of George Washington University Medical School in Washington, D.C. Throughout the 1950s, Watts had performed numerous lobotomies on patients diagnosed with depressive disorders and schizophrenia. The most common operation, which Watts developed with psychiatrist Walter Freeman, was called a "transorbital lobotomy," in which an implement resembling an ice pick was inserted through the patient's eye sockets into the frontal brain lobes. This procedure aimed to sever the connections between prefrontal lobes and those deeper limbic structures assumed to be the seat of emotion. In theory such a disconnection would eliminate the emotional outbursts in a number of psychiatric conditions.[37]

In July 1955 Watts performed the first recorded transorbital lobotomy on a ticcing patient, a thirty-seven-year-old male who had incapacitating motor tics, coprolalia, echolalia, and echopraxia. The patient had been forced to leave school at fifteen and since then had remained at his parents' home. Watts was persuaded to attempt the lobotomy because the patient's tics continued to increase in severity. Watts reported that the patient was more relaxed after the operation, even though "his mannerisms persisted although with less emphasis and force." Both the patient and his family were pleased with the results. For the first time the patient could join in social events without embarrassing his mother and others. He slept without making barking noises and was generally less tense and restless. Watts's follow-up two years later found a "decrease in the frequency, duration, and amplitude of the motor tics and the compulsive swearing." Ten years later, family members reported that the patient's jumping and jerking had disappeared and they described him as "calm"; he only cursed when excited.[38] Watts concluded that transorbital lobotomy was appropriate for intractable ticcers on whom all other interventions had failed.

In 1962, Toronto psychiatrist E. F. W. Baker, citing an experiment in which involuntary movements resulted from lesions made in a monkey's midbrain, wrote that he was persuaded "that some such lesion may underlie the Gilles de la Tourette's syndrome." Conceding that there was a sizable gulf between these findings and his twenty-two-year-old patient's severe tics, coprolalia, and panic attacks, Baker nevertheless decided to perform a lobotomy, whose aim was to override the putative lesions that had caused the ticcing, cursing, and panic. According to Baker, the surgical lesions would correct the abnormal signals between the prefrontal lobe and the thalamus, a structure deep in the brain, and reduce the patient's panic attacks and tics. He explained that "the decision to perform leucotomy [lobotomy] in this case hinged on . . . how much reduction of tension could be expected, at the expense of how much loss of 'control.'" Baker claimed that as a result of the operation the patient's tics and panic attacks were significantly reduced, and that two months later the motor tics were reduced to two each half-hour. However, Baker reported "an unfortunate postoperative complication of staphylococcal frontal lobe abscess with grand mal seizures." A combination of antibiotics and maintenance anticonvulsants was used to treat the infection and regulate the seizures. A year after the operation Baker found that the patient's tics and panic attacks were still in remission. The patient attended night school regularly, no longer had self-destructive fantasies, and had hopes for a "reasonable future." Yet since the operation, the man exhibited "a moderately short attention span and a dearth of ordinary social contacts." Baker concluded that although "the patient is 'socially acceptable,' . . . it is, of course, too early to make a final assessment of the value of the treatment in this case."[39]

Two decades later, German (Hamburg) child psychiatrists Uta

Asam and W. Karrass published a follow-up study of sixteen Gilles de la Tourette's patients (two of their own and fourteen others, including Baker's patient) who had been treated with psychosurgery. They concluded that "neurosurgery in Gilles de la Tourette's disease involves considerable risks." Most patients suffered from "severe side effects" and their general condition often "deteriorated" after the procedure. Only "if the observation time was short" could the surgery be characterized as effective. Immediately after the operation most patients' symptoms appeared to ameliorate, but after two years the overwhelming majority reported a return of symptoms that in most cases were more severe than those prior to the surgery. As troubling was the discovery that half of the patients experienced permanent postoperative neurological damage including spasticity, stuttering, brain infections, difficulty with balance and walking, as well as general cognitive deficits. Although three of the patients appeared to have positive personality improvements, four developed new psychosocial difficulties after the operation. In sum, only five of the sixteen could be considered to have met any criteria of improvement, while the remaining eleven appeared worse after their operations than before.[40]

Chlorpromazine

Introduced in Europe in 1952 and in the United State three years later, the neuroleptic drug chlorpromazine (Thorazine) immediately proved to be an extremely effective agent for the control of schizophrenic symptoms. Its impact seemed to be almost miraculous; patients who had been institutionalized for years were suddenly able to rejoin their families and to return to productive lives. Within a few years of its introduction chlorpromazine was prescribed for other psychiatric disorders, including major depressions, anxiety, and hyperactive mania.[41] By the mid-1950s chlorpromazine was routinely administered to patients as a management drug under the assumption that its calming effect made other procedures, such as carbon dioxide inhalations and psychosurgery, more effective.

In his 1957 article reporting success in controlling motor tics and coprolalia with CO_2 inhalations, Michael noted in an aside that his patient was also given twenty-five milligrams of chlorpromazine twice daily for ten days.[42] MacDonald, who also used carbon dioxide treatments, was convinced that chlorpromazine had played a central role in controlling his patient's symptoms so that the inhalation therapy could be effective.[43] Similarly, Watts, reporting on the success of a transorbital lobotomy in relieving severe symptoms, stated without comment or explanation that after the operation the patient was placed on a regimen of 100 milligrams per day of chlorpromazine.[44] Finally, Baker had discharged his patient "on a regimen of 600–800 mg. chlorpromazine daily."[45]

Although these physicians had paid little attention to chlorpromazine's possible impact on their patients' tics, others soon made the connection. High dosages of chlorpromazine produced Parkinsonian symptoms, or rigidity of movement, because the drug reduces the transmission of the neurotransmitter dopamine, resulting in the opposite symptoms from those displayed by ticcers. Thus it was reasonable to assume that chlorpromazine might be effective in reducing involuntary tics, which many observers believed resulted from hypertransmission of dopamine.

Citing Michael's report, British psychiatrist S. Bockner treated two ticcing patients with chlorpromazine in 1959. The first, a sixty-five-year-old male who had facial tics at age five and vocal tics by his teen years, had spent his life under the sway of waxing and waning uncontrollable sounds, swearing, and unwanted movements. The patient had been subjected over the years to a variety of unsuccessful interventions, including a tracheotomy to reduce his vocalizations. Bockner decided to place the patient on 150 to 225 milligrams daily of chlorpromazine. "There was," according to Bockner, "a rapid and remarkable response . . . The motor spasms became much less frequent and the vocal tics almost completely disappeared." However, a second male patient, sixty-seven years old, who also showed a positive response, developed jaundice and the drug was discontinued.[46] Nevertheless, these results convinced Bockner that chlorpromazine "should be the first line of treatment." Unlike others, Bockner did not feel compelled to defer to psychoanalytic theory, noting instead that "the relationship of Tourette's disease to chorea is interesting and important in that it suggests that there may be an organic basis to the condition." He also pointed out that others had reported that tics often occurred during or following encephalitis and suggested that "the vocal tic of Tourette's disease involves the respiratory musculature in a very similar manner to these post-encephalitic tics."[47]

Bockner's views, however, were unique. More typical was the

reaction of the American psychiatrist Alvin M. Mesnikoff, who reported in 1959 on three "schizophrenic" tic cases, none of which improved under psychoanalysis alone, but only with the addition of chlorpromazine. However, Mesnikoff argued that the chlorpromazine only served to help calm the patients so that the psychoanalysis could be invoked.[48]

That same year, three of the most influential American psychiatrists, Leon Eisenberg, Eduard Ascher, and Leo Kanner, all from Johns Hopkins University Medical School, collaborated on a review essay about the current knowledge of tics for the *Journal of American Psychiatry*. Claiming that "pharmacologic therapy, including sedatives, muscle relaxants, and tranquilizers, has been ineffective," the trio nonetheless conceded that the outcome of psychotherapeutic treatment of tics was uniformly poor, often resulting in insanity. Captives of the assumptions of psychoanalysis, Eisenberg, Ascher, and Kanner could conceive of no alternative to psychoanalytic therapy. They concluded that "it would appear justified to continue efforts at psychotherapeutic management despite" the "admonition that the tic phenomena are defenses against impending psychosis and Mahler and Luke's warning that the release of aggressive erotic material 'may weaken the controlling powers of the child tiqueur.'" At the very least, they argued, "psychotherapy may be of assistance to child and family in dealing with the social consequences of the illness."[49]

Eisenberg, Ascher, and Kanner's dismissal of pharmacologic therapy for ticcing patients resulted in a year-and-a-half-long debate in the letters to the editor of the *American Journal of Psychiatry*. What these letters revealed most of all was the extent to which even advocates of pharmacological interventions felt compelled to defer to psychoanalytic assumptions. The exchange exposed more about professional power relationships than it did about therapeutic efficacy.[50]

Europeans also were experimenting with chlorpromazine for the treatment of tics, but seemed less compelled to attempt to interpret their successes as confirmation of Freudian insights.[51] In 1962 two Polish psychiatrists, Roman Dolmierski and Maria Kloss, published a comprehensive essay entitled "de la Maladie de Gilles de la Tourette," in the French journal *Annales Médico-Psychologiques*.[52] Tracing the history of the diagnosis from Gilles de la Tourette's 1885 paper, they focused on the work of Clauss and Balthasar in the 1950s and others who had "suggested that obsessive neurosis and the tics of Gilles de la Tourette were both conditions that were caused by brain lesions."[53]

As evidence, Dolmierski and Kloss cited two cases (out of five) that they had treated with chlorpromazine in the last ten years in their clinic in Gdansk. The first, a twelve-year-old boy with multiple motor tics, coprolalia, and echolalia, had a history of infections. At twelve months, following a vaccination, he had a fever of 40° C (104° F), with two convulsive crises. Soon after, the boy contracted measles. At seven years, his tics began, followed by coprolalia. Placed on chlorpromazine (225 milligrams per day), the tics and coprolalia diminished to eyebrow movements, slight facial grimaces, and slight muscle movement of the shoulders and neck. The second patient, a fifty-five-year-old woman with neck and shoulder tics and coprolalia, experienced her first symptoms in early childhood. By the time she was fifteen the tics worsened and she developed coprolalia and obsessive symptoms. Placed on a dose of 100 milligrams of chlorpromazine per day, her symptoms were controlled but not eliminated.[54]

Based on these results, Dolmierski and Kloss concluded that Tourette's disease was most likely organic. In both cases "the results of the treatment with chlorpromazine seemed to affirm these hypotheses" because "it is known that chlorpromazine acts on the spinal cord and on the sub-cortical neurons and that in high doses results in Parkinsonian symptoms." Moreover, in both cases this medication reduced or eliminated tics, but when it was withdrawn, the tics reappeared.[55]

Haloperidol

Like chlorpromazine, the antipsychotic drug haloperidol (Haldol) had been widely used first in Europe during the 1950s for the treatment of schizophrenia. As is now known, unlike chlorpromazine, whose action inhibits dopamine transmission throughout the brain, haloperidol appeared to act exclusively on the transmission of dopamine in that part of the brain that regulates the movements most associated with tics, the substantia nigra to the basal ganglia circuit.[56] Haloperidol's tranquillizing effect in reducing dopamine transmission made it a particularly effective drug for control of violent or otherwise unmanageable psychiatric patients.

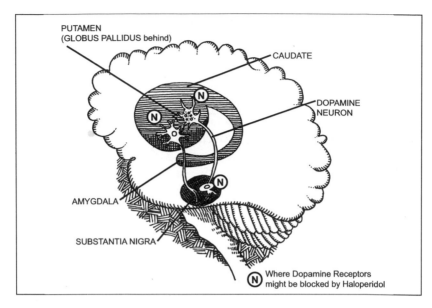

Figure 8.2 Neurons connect the basal ganglia with the dopamine-producing cells of the substantia nigra. Hypersensitivity of these neuron receptors to the neurotransmitter dopamine has been suggested as a major reason for the disruption of the basal ganglia circuitry and the resultant involuntary movements of Tourette's. The neuroleptic drug haloperidol binds to dopamine receptors, reducing the dopamine transmission and thus suppressing the involuntary movements.

The first report of success with haloperidol in treating tics appeared in an article published in 1961 by the French psychiatrist M. Jean N. Seignot. A thirty-five-year-old man who, according to Seignot, "presented with typical maladie des tics" had, since he was ten, submitted to every type of treatment, including a lobotomy ("topectomie") in 1957, but with no amelioration of symptoms. By November 1960, the man's motor tics and vocalizations were alarming, averaging one or two tics per minute; each tic lasted around four seconds, during which the patient repeatedly hit himself, causing a series of hemorrhages and blindness. The patient also had coprolalia and frequently mixed his cursing with uncontrolled insults. Moreover, he "was constantly obsessed by an impulse to hit his forehead against the sharp edges of furniture and doors." Because Seignot was familiar with the success of haloperidol in managing violent outbursts, he decided to prescribe it, then known by its experimental label R.1625.[57] Results were immediate and, wrote Seignot, "remarkable." "At a dosage of .6 milligrams per day, the patient's tics were reduced to only two to three per day, moreover, the intensity of these residual manifestations was extremely attenuated; in particular the coprolalia completely disappeared."[58] Three months later, at the time Seignot's article was submitted, the patient's tics still had not returned, even though he suffered from the additional stress of illness in his family.[59] Although Seignot conceded that he could not generalize from his experience with one admittedly unusual case, he hoped that his results with R.1625 would lead others to use it "not only on this grave form of tics, but also in those other, more ordinary and much more widespread cases found among adults and even children."[60]

"Seignot's encouraging report," wrote Iowa State University psychiatrists George Challas and William Brauer, "prompted us to try haloperidol on two cases of Tourette's disease." Obtaining a supply of the still experimental drug from McNeil Laboratories in 1962, Challas and Brauer selected two pediatric male patients for their trial. One, a sixteen-year-old male who had first developed neck and facial tics along with coprolalia when he was seven, had been treated over the years with the usual range of therapies—psychoanalysis, hypnosis, behavioral therapy, medications (including chlorpromazine), and electroshock—none of which had any lasting effect. When Challas and Brauer saw the boy in March 1963, he presented with "frequent facial tics, jerking of the arms, spitting, obscene gestures, and shouts of 'fuck' approximately three times per minute." Placed on 1.5 milligrams of haloperidol for nine weeks, the boy's coprolalia and motor tics immediately decreased in frequency and intensity. The boy's abnormal liver function, however, caused Challas and Brauer to suspend this treatment, and a few days later the motor tics and coprolalia reappeared with the same intensity and frequency as before. The second patient, who in April 1963 was ten years old, displayed similar symptoms to the first and had gone through parallel treatments with the same negative result. Again, the psychiatrists administered haloperidol and the patient's tics and coprolalia almost immediately disappeared. With no marked or persistent side effects, as in the first case, the second child was still receiving the drug when the article was submitted in 1963. "The palliative effect of R1625 in the above classic cases of Tourette's disease," concluded Challas and Brauer, "was dramatic and closely parallels Seignot's experience." They were convinced that haloperidol was "worthy of trial in cases of severe tics as in Tourette's disease."[61]

Challas and Brauer continued to follow their two patients. The next year, they reported that because their first patient's liver was functioning normally, they had reinstituted haloperidol treatment. "At time of writing," they reported, the "patient is on a maintenance dose of 3.2 milligrams of haloperidol daily, shows no side effects and feels that in spite of an occasional explosive (but subdued) bark, he can function well in his community." The second patient "continued to do well on this medication and at time of this writing is symptom free and has graduated into Junior High School with good grades." His maintenance dose was 1.5 milligrams daily.[62] Meanwhile, three of Challas and Brauer's Iowa colleagues, James L. Chapel, Noel Brown, and Richard L. Jenkins, reported the cases of two additional ticcing patients treated successfully with haloperidol.[63] Yet Chapel, Brown, and Jenkins continued to defer to their powerful psychoanalyst colleagues. "It should be obvious," they concluded, "that the problem of Tourette's disease is not resolved by control of the tics through medication." The trio seemed almost apologetic about their success with psychopharmacology. "However, in view of the extent to which coprolalic tics provoke social rejection," they wrote, "the difficulties in effective psychotherapy with individuals with uncontrolled coprolalic tics are extreme. We regard the control of this troublesome symptom as the greatest importance in creating a situation more favorable for effective readjustment of the patient."[64] Perhaps these words were the conventional and deferential price one had to pay to obtain publication of results that could be seen as undercutting psychoanalytic dogma. The next year, when doctors Arthur and Elaine Shapiro submitted similar results to these same journals, but in language that attacked rather than deferred to psychoanalytic insights, their article was rejected.[65]

In the same issue of the *American Journal of Psychiatry* in which Chapel, Brown, and Jenkins's article appeared, a Michigan psychiatrist, Alexander R. Lucas, reported that two of his ticcing and coprolalic patients "have shown definite and long-lasting improvement of their tics" with two powerful antipsychotic phenothiazine drugs (triflupromazine and thioridazine [Mellaril]), which today are known to work generally on reducing dopamine transmission.

Like Chapel and colleagues, Lucas was reluctant to find fault with psychoanalysis or to claim that the success of medication pointed to an organic pathogenesis of tics. "In these patients," wrote Lucas, "the improvement of coprolalia, barking tics and body tics seemed directly related to the administration of the two phenothiazine drugs used and resembled the results reported with haloperidol." But, Lucas conceded, both patients "had many associated psychological problems, and these cases emphasize the need for overall therapy, in addition to the drug treatment." He concluded that "the phenothiazine drugs may be very useful adjuncts in the treatment of tic syndrome," which, nevertheless, remained at bottom psychogenic.[66]

Meanwhile in Europe a number of publications appeared affirming reports on the effectiveness of haloperidol. Unlike their American counterparts, none of these psychiatrists felt obliged to defer to psychoanalytic assumptions in their reports. For instance, in 1964 two Italian psychiatrists, U. Perini and B. Lampo, reviewed thirty cases, concluding that treatment of tics "with haloperidol constitutes a therapy which can be considered elective and advisable," even though they believed that it controlled rather than cured these symptoms.[67] That same year, a French physician, T. Sobierski, reported that his clinical results in "numerous cases" confirmed that haloperidol produced an effect on the brain that convinced him that it acted on the causes of Gilles de la Tourette's disease.[68]

The next year Diane H. Kelman, a British clinical psychologist at Maudsley Hospital in London, published a summary of forty-four cases of Tourette syndrome that had appeared in medical journal articles over the past sixty years to see if she could determine which treatments appeared to result in the best outcomes.[69] Again, the success of haloperidol in controlling tics and coprolalia was seen as compelling evidence to rethink the etiology of multiple motor tics and coprolalia. Compared to all other interventions—including psychotherapy, hypnosis, narcoanalysis, carbon dioxide inhalation, and electroconvulsive shock—the "phenothiazines (chlorpromazine, triflupromazine) and piperazines (trifluoperazine, thioridazine and haloperidol) were more than twice as likely to result in symptom improvement."[70]

The most convincing and complete report of the effect of haloperidol on a single patient appeared December 1966. Neurologist Janice R. Stevens and psychiatrist Paul H. Blachly, at the University of Oregon Medical School, performed the first single-blind study, demonstrating by temporary placebo substitution that haloperidol's action inhibited these tics. In addition, Stevens and Blachly provided practitioners with an explanation of the likely chemical mechanisms of the drug on neurotransmission.[71] Yet, Stevens and Blachly's work is rarely acknowledged in the current vast literature on Tourette syndrome. It is, therefore, instructive to look closely at what they wrote to see why this may be so.

Stevens and Blachly's thirteen-year-old female patient had developed moderately severe twitching in her left arm three years before. She presented with convulsive symptoms initially diagnosed as "seizures." Treated with a variety of unsuccessful interventions, including "cervical spine manipulations" from a chiropractor, the girl's tics increased. In addition, another physician diagnosed the symptoms as "St. Vitus dance," but suggested no specific treatment. The parents were convinced that the tics resulted from their attempts to discipline their daughter because the movements "could often be terminated by slapping the child violently." By the time she was admitted to the University of Oregon Hospital in 1964, the girl was never tic free for more than two minutes and often suffered "continuous irregular violent jerks of the head, neck, and shoulders which occurred 10 to 15 times per minute." Aversive therapy was attempted during which the patient was subjected to "a systematic program of shock and verbal reinforcement (patient saying the word "shock" to each tic) during hour-long weekly sessions." Although these resulted in a reduction of the frequency of tics "in the laboratory setting," there was no "carry-over to the school or home environment." After "six months of weekly conditioning treatment with shock plus interim verbal reinforcement, insight therapy, relaxation therapy, and hypnosis, the patient was distinctly worse, having added a loud hissing through her teeth 10 to 15 times each minute to the motor tic." When placed in Stevens and Blachly's care, the patient was given a series of drugs, including chlorpromazine, primidone, and dextroamphetamine, but these only irritated her and had no effect on her tics. Frustrated, Stevens and Blachly then learned about Challas and Brauer's report in the *American Journal of Psychiatry*. They instituted a dosage of 2.5 milligrams of haloperidol four times daily. The girl's tics and hissing dramatically decreased, and "at 3 mg four times daily both motor tics and hissing entirely disappeared." In addition, her parents reported "an extraordinary" improvement in their daughter's behavior. She was now "a cooperative, friendly teen-ager who was careful in her appearance and improved dramatically in her home- and schoolwork."[72]

Having been led astray before by the patient's temporary improvements, the team decided to initiate a clinical trial during which they substituted a placebo for haloperidol without informing the patient. The result was dramatic: the girl's twitching returned three days later, and by the sixth day her tics reached their pretreatment severity. Three days after the reinstitution of haloperidol treatment, the tics were again under control. The patient received 2.5 milligrams of haloperidol for the next ten months without recurrence of the symptoms. At this point Stevens and Blachly submitted their article for publication.[73]

Like their predecessors, Stevens and Blachly had assumed that their patient's symptoms were "explicable in functional [psychoanalytic] terms" because she "seemed to be escaping both expectations at school and discipline at home by the bizarre spasmodic tic which was regularly exaggerated during stress or punishment." But after their clinical trial with haloperidol, they were persuaded that "the underlying cause . . . is indubitably organic." Stevens and Blachly compared their patient's ability to temporarily inhibit her tics while "under aversive constraints" to patients with advanced Parkinson's whose characteristic tremors and rigidity could sometimes be dramatically and temporarily overcome when they "run with considerable agility to the window to witness an exciting spectacle, dance to infectious music, . . . or surmount paralysis of volitional movements by mimicry of others." However, this temporary recovery "does not prove that a malady is hysterical or 'nonorganic' in origin." Rather, wrote Stevens and Blachly, this result was due to "immediate strong reinforcements or rewards that may mobilize facilitatory or inhibitory mechanisms unavailable during ordinary automatic or volitional activity." This reinforced the authors' belief that aversive therapy could not control behavior outside of the laboratory. "Indeed," wrote Stevens and Blachly, "the difficulty of behavior techniques and the dramatic improvement induced by only one of a number of potent pharmacological agents" provided powerful evidence that refuted an exclusively psychological cause of la maladie des tics. They went even further, speculating that, given its chemical similarity to the "neuroinhibitory transmitter, μ-aminobutyric acid (GABA)," haloperidol worked to decrease dopamine transmission, and thus reduced the motor activity signals sent out by the brain's basal ganglia.[74]

Stevens and Blachly emphasized that both clinical experience and their understanding of biochemistry provided persuasive evidence of the organic etiology of tics. Nevertheless, in conceding "that environmental factors modulate an underlying physiologic dysfunction," they preserved a role for psychosomatic features in the pathogenesis of the disorder.[75] Yet Stevens and Blachly unambiguously believed that pharmacological agents should be the therapy of choice, and that psychodynamic and behavioral psychiatry must now be placed in a supporting role in the treatment of tics.

Still, the effectiveness of haloperidol could be read by others as undergirding rather than undercutting psychoanalytic claims. In August 1967, Challas, Chapel, and Jenkins, three of the pioneers of the use of haloperidol on ticcing patients, published a follow-up of eleven patients, a number of whom remained tic free on low maintenance dosages of haloperidol. "We are deeply impressed with the control of symptoms," they wrote, "made possible by the use of haloperidol." The six patients who remained on haloperidol, for periods of time ranging up to three and a half years, all reported that their symptoms had effectively abated. "The response of patients with Gilles de la Tourette's disease to haloperidol," wrote Challas, Chapel, and Jenkins, "is prompt and dramatic." Nevertheless, the trio disagreed with Stevens and Blachly's assertion that the etiology of Tourette's was organic. "Some authors believe an organic disturbance of the brain underlies the symptoms. However," wrote Challas, Chapel, and Jenkins, "neuropathologic evidence is lacking." Rather, they "believe[d] that psychologic factors, perhaps acting on an organic substratum, are responsible for the symptoms." Some of the patients were "controlling and punitive," while others had displayed "a tendency to react with over-indulgence and overprotectiveness." Challas and his colleagues were convinced that "psychodynamically the condition seems to develop under conditions encouraging repression."[76] As Chapel had explained a year earlier, success with haloperidol suggests that "this syndrome may be the result of an organic brain abnormality which is provoked by psychic stress. A psychological component seems clear, however, and this component is very similar to a conversion reaction" in which "intense anger and hostility is directed toward

one or another parent but cannot be expressed in the open voluntary verbal or behavioural fashion because of the child's insecurity
with the parent."[77] Yet, Challas, Chapel, and Jenkins all admitted
that unlike haloperidol, which "has been consistently successful,
in our hands, in controlling the symptoms of Tourette's disease,
. . . psychotherapy measures have appeared beneficial in some cases
but have rarely removed the symptomatology completely."[78]

Such logic was not limited to North American practitioners. In
1967 two Finnish pediatricians reported success with haloperidol
in managing the symptoms of a boy diagnosed with Gilles de la
Tourette's syndrome. Nevertheless, the authors still viewed the illness as psychogenic: "It is possible," they wrote, "that this condition is specifically connected with repressed aggressive emotions
. . . The lessening of aggressions might lead to the disappearance
of symptoms. The improvement attributed to haloperidol may possibly also be connected with this, as haloperidol is usually indicated
in aggressive manic and psychomotor agitation."[79]

By 1968, consensus had developed among North American psychiatrists and neurologists that haloperidol was an extremely effective agent for the management of motor tics and coprolalia. The
most influential voices, however, continued to accept a modified
version of Mahler's belief that even if tics had an organic substrate,
their underlying enabling factors were psychodynamic. A group of
neurologists and physiologists at the Mayo Clinic reported in 1968
that their successful use of haloperidol on a patient "reinforced
our belief that this disorder is the result of organic changes within
the central nervous system," but they were convinced that it was
"obvious" that "psychological factors are important, in both the
development of the behavior and personality changes and the treatment of Tourette's syndrome."[80]

As a result of such arguments by those who had reported clinical
success with haloperidol, most North American psychiatrists in
1968 continued to view their patients' tics and coprolalia in the
context first authorized by Margaret Mahler a quarter of a century
before. Typical was the 1968 report published in the *Journal of
Nervous and Mental Disease* by psychiatrists Hilde Bruch and
Lawrence C. Thum from New York State Psychiatric Institute,
where Mahler had done her clinical work on tics. Bruch and Thum
connected the severe tics and coprolalia of a twelve-year-old boy
named Harry to a "maternal psychosis."[81] "From the moment of

the child's conception, the mother confessed, she had considered him a reincarnation of her own evil and aggressive impulses. She had made control of his aggression the goal of her life, rearing him with extreme psychic restraint." According to Bruch and Thum, "the mother directly expressed the dynamic constellation that has been recognized by others as characteristic of maladie des tics (without encephalopathy). The old literature emphasized a close association between maladie des tics and schizophrenia. The early development of this patient was not incompatible," wrote Bruch and Thum, "with a later schizophrenic reaction" brought on by the mother's hypercontrol. Ten months of psychoanalytic treatment of Harry and his mother (separately) brought about a "dramatic change of the patterns of intrafamilial interaction" and "permitted" the boy's tics to ameliorate. Rejecting an organic etiology, Bruch and Thum returned to Mahler's "extensive psychoanalytic studies of the tic syndrome" in which she "spoke of 'the mother and child interdependence as having been quite extreme at times.'" They endorsed the belief that antipsychotic drugs should be used to break down resistance to psychotherapy in order to expose the "'maternal dominance' . . . characteristic of the families of . . . tic patients."[82]

Neither the numerous anomalies in psychoanalytic treatment nor even the discovery of the effectiveness of haloperidol for the management of tics had much immediate persuasive power on American psychiatry. This is not surprising. Few professions historically have demonstrated a willingness to accept approaches and evidence that undercut their intellectual and political assumptions. Rather, like the psychoanalytic reaction to the effectiveness of haloperidol, professions attempt to integrate seeming contradictions into their worldview. Psychoanalysts themselves proved extremely resistant to the claims of the success of treatments with antipsychotic drugs. Even into the 1980s, long after the effectiveness of dopamine suppressors in regulating tics had been established, psychoanalytic spokespersons continued to insist that it would be a mistake to confuse a symptom (motor and vocal tics) with the underlying psychological forces that enabled it.[83] Such a result, according to this view, proved nothing about pathogenesis and only demonstrated what was obvious: the manifestations of psychogenic tics were physiological and, given a powerful enough tranquillizing agent, these symptoms could be suppressed.

In contrast to the prewar years, during the postwar period there was no scarcity of medical journal reports of ticcing and cursing patients. Whether this represented an actual increase in the afflicted, in the reporting, or increased hope by those with tics that medicine might offer better interventions is difficult to determine. There was certainly no lack of claims by practitioners of the effectiveness of their latest interventions. Even though most practitioners, especially those in Britain and North America, attempted to fit successful somatic and organic interventions into the psychoanalytic paradigm, they had inadvertently exposed the weaknesses and contradictions of the psychoanalytic frame. However, this alone does not explain the triumph of the organic frame in Britain and North America. Contrary to a pervasive view among historians of science, anomalies did not seem to account for the shift from psychoanalytic psychiatry.[84] Rather, elements outside of psychiatry provided the impetus to rethink the psychoanalytic paradigm, especially when it came to Tourette syndrome.

Before examining this story, it is important to return to France. Especially in Paris, where more than a century ago Georges Gilles de la Tourette first identified the involuntary ticcing and cursing as a unitary illness, Tourette syndrome is rarely acknowledged and, when it is, it is treated almost always as a psychogenic condition. In France the psychoanalytic frame grew in power in the 1960s and 1970s at the very time when the persuasive evidence increasingly appeared, much of it from French researchers, for an organic substrate for convulsive tics. By the 1990s psychoanalysis still dominated the diagnosis and treatment of multiple motor tic and coprolalia. It is crucial to examine why psychoanalysis continues to hold sway in France if we are to understand fully why it is no longer taken seriously in Britain and North America.

The French Resistance

$$\cdots \; 9 \; \cdots$$

In 1941, during the Nazis' occupation of Paris, two French psychiatrists, at separate meetings, presented very different explanations for the etiology of maladie des tics. The circumstances of each presentation would prove to be as influential as the substance of each report. The first paper, which argued that tics had an organic etiology, was presented at a meeting of the Société Médico-Psychologique held in Nazi-occupied Paris in February with the tacit approval of the Vichy authorities. Later that year it was published in *Annales Médico Psychologiques*.[1] The second paper, which took a psychoanalytic approach, was delivered clandestinely in April to a group of underground psychoanalysts. This paper would not be published until six years later in the journal *L'evolution Psychiatrique*,[2] reestablished by French and émigré survivors of the Nazi and Vichy purges of psychoanalysts, many of whom were of Jewish origin.

Jean Delay, author of the first paper, was appointed in 1942 to head the University Psychiatric Clinic Sainte-Anne in Nazi-occupied Paris.[3] After the war (in 1951) Delay and his colleague Pierre Deniker were the first to treat schizophrenics successfully with chlorpromazine (Thorazine).[4] In his 1941 paper, Delay argued that tics should be placed in the same context as other involuntary movements, such as chorea, epilepsy, tremors, and tetanus, which slowly had "passed from the category of neuroses" to "those of organic disorders of the central nervous system" caused by a "specific lesion of the brain."[5]

In support of this general proposition, Delay claimed that Henry Meige, at the end of his life, had changed his mind and decided that the majority of tics resulted from lesions in the midbrain. Delay quoted from an uncited report in which Meige purportedly revealed that unanticipated neurobiological findings had persuaded him to renounce his earlier psychological views.[6]

In contrast, wrote Delay, psychoanalysts continued "seeing in the very expression of the tic an unconscious motivation, insisting on the *subjective* character that cloaks the tic (for example, when it replaces a child's masturbation or enuresis [bedwetting]) and on the neurotic tendencies of its expression." Delay, however, insisted that even a tic that appeared to be a reaction to a psychological conflict was, nevertheless, a sign of an organic brain lesion. The appearance of a tic after an emotional experience, wrote Delay, was often only a coincidence: the "'intentional' character of a tic is always contestable." Although "the association of a psychopathic state is often [found] in organic tics, . . . it must not be forgotten" that an injury to the midbrain can "frequently" result "in emotional troubles," especially if there are "lesions of the grey matter of the brain."[7]

As evidence for his argument Delay cited the case of André R., a twenty-three-year-old soldier who initially was diagnosed as having neck tics of a "neurotic" origin. The tics first appeared a month after André witnessed the death of a friend in an automobile accident. André's other symptoms included insomnia, anxiety, sensitivity to noise, and "black thoughts." Even bringing up the subject of tics with André was enough to set them off. André had always been extremely emotional and his tics were accompanied by obsessional preoccupations that increased his anxiety. An examination of André's long-term medical history, however, revealed that he also had a history of sensory problems including numbness and tingling, hot and cold sensations on his neck and head, and feelings of an electric discharge in his arms and legs. Moreover, four years previously André had a neck tic, similar to the one he now had, which lasted about three weeks and "frightened him a great deal" at the time. Three years later, about a year before his current symptoms, André experienced a temporary rapid decline of eyesight after an eye irritation, which was diagnosed as "transitory amblyopia."[8] Given André's longtime symptomatic history, Delay was extremely skeptical of the diagnosis of mental tics. Rather, André's history of neurological symptoms, when combined with a neurological examination, convinced Delay that André's proper diagnosis should be multiple sclerosis.[9]

Delay's point was not that multiple sclerosis was often misdiagnosed as psychogenic tics. Instead, Delay was convinced that a variety of different neurological conditions, including encephalitis, lesions in the midbrain, or strokes, might lead to involuntary movements. He chose André's case as a demonstration of how an emphasis on a patient's mental state, combined with an incomplete neurological perspective, might persuade an unwary clinician to make a diagnosis of mental tics. Organically caused tics, wrote Delay, "can present themselves with all the characteristics of mental tics," and a patient's "obsession" with and "anxiety" about his tics can persuade an observer of their psychopathological origin.

In contrast to Delay, French psychoanalyst Julien Rouart, who had met secretly with his colleagues in April 1941, argued that convulsive tics were psychogenic in character. Like the French psychoanalysts who would follow him, Rouart was influenced both by Freud and by Freud's French rival, the Salpêtrière-trained psychiatrist Pierre Janet (1859–1947), who in the late 1890s had published extensive case histories of ticing patients.[10] Unlike Freud, Janet did not immediately develop a school of followers. Even though Janet, in contrast to Freud, had examined, analyzed, and published numerous cases of ticing patients, his views on tics had proven much less influential than those of Freud. Nevertheless, Janet had trained a cadre of influential psychiatrists who, after the Second World War, emerged as the leaders of French psychoanalysis. This group, including Rouart and Serge Lebovici and his students, rescued Janet's work on tics and made it the centerpiece of what emerged as the French psychoanalytic view of ticing behaviors.

In contrast to Freud's concept of the repressed unconscious, Janet spoke of the subconscious—a cluster of autonomous and unregulated ideas, memories, and fantasies that were disconnected, or split off from consciousness. Because Janet's subconscious was not actively concealed and disguised in the same way as Freud's repressed unconscious, it was, in Janet's theory, more directly accessible and more easily modifiable by hypnosis and suggestion.[11]

Janet's construction of the subconscious had important implications for his explanations of ticing behaviors, whose causes, he thought, were accessible to the afflicted and curable by clinicians.

Janet believed that many tics evolved from earlier voluntary actions that later developed into repetitive habits. "A schoolgirl was dissatisfied with the place allotted to her in the schoolroom, and pretended that she felt a draft on her neck coming from the window." She would elevate one of her "shoulders as if to bring her clothes a little more closely round her neck." Then, reported Janet, "she commenced to depress her head and to indicate her displeasure by facial grimaces." These behaviors, which began as a childish attempt to impose her will, "eventually passed beyond voluntary control."[12]

Building on and elaborating observations first made by Guinon in the 1880s, Janet was convinced most tics were symptoms or signs of subconscious obsessive-compulsive conflicts (what Janet called "psychasthenia").[13] A child's tics, wrote Janet, were the "signal symptom of an obsessional psychoneurosis."[14] According to Janet, afflicted children were tormented by an inner sense of imperfection, arising from disappointment with their own performance or lack of a sought-for satisfaction. Janet believed that this dissatisfaction was at the root of all tic symptoms. "Tics are systematized but useless movements," but because of their symbolic meaning, they were open to psychological analysis.[15] As an example Janet reported the case of a thirty-year-old man with tics who exhibited extreme perfectionist behaviors that made it impossible for him to accomplish simple tasks, such as writing two letters of the alphabet consecutively. Train travel produced feelings of extreme agony.[16] Janet traced the patient's anxieties and perfectionism back to childhood.[17] Yet, because the man's perfectionism was subconscious, Janet was certain that tics such as these could be relieved by removal of the patient from psychologically stressful situations.[18]

Drawing on Janet to interpret the symptoms and case histories of his own patients, Rouart reached three main conclusions. First, like Janet, Rouart believed that a tic was a compulsion that was almost an obsession. But, like Freud, Rouart also was persuaded that a tic often assumed "the structure of hysteria by its symbolic hypermimic character." Second, and again affirming Janet, Rouart found that observation of child ticcers always revealed the same character traits, including emotional retardation, antagonistic behavior, unremitting aggressiveness, neurotic family conflicts, and marital discord between the parents. In short, a tic was "situated in a personality and in the entire family." Third, affirming Janet's central dictum, Rouart argued that tics were "filled with *significance,* but this significance generally is not immediately apparent," because of the tics' contradictory functions. The task of the therapist was to uncover the hidden and contradictory meanings of these symbolic gestures to shed light on the ticcer's repressed familial conflicts. This was often difficult because the physical manifestation of the tics was always "incomplete" and "secretive." For instance, tics "can appear as a *substitution* for a forbidden act or one that is impossible to achieve." But a tic also always revealed a person's conflicts about developmental issues as she or he interacted with the complexities of relations within the family. As with other manifestations of profound psychological conflicts, Rouart believed that only psychoanalysis could untangle the multiple and contradictory meanings of tics.[19]

As evidence for the psychogenic origin of tics, Rouart noted that tics frequently changed their location, replacing one another according to certain external conditions. He pointed to the case of nine-year-old Guy B., whose tics altered from facial, to pursing his lips, to wrinkling his nose to the left, to blinking, to inclining and shaking his head to the right. According to Rouart, whenever Guy was reprimanded by members of his family for displaying certain tics, he would suppress them, only to have others appear in their place. This dynamic convinced Rouart of the wisdom of Janet's suggestion that tics could be relieved by removal of the child from stressful situations. Additional evidence that tics were "emotionally provoked" was found in the case of a twelve-year-old patient whose multiple motor tics were exacerbated at school, but disappeared during vacations.[20]

According to Rouart, obsession, as "described by Janet," was the most important psychological issue found in ticcers. Rouart quoted one child who, like many others, explained that he felt "obligated" to tic. Tied to this obsession was a sense of shame. "Among adults or older children, consciousness of ticcing is more frequent and the sense of shame can give way to all sorts of processes leading to deceits." Rouart returned to Meige and Feindel's report of the "Confessions of a Tiqueur," noting that "O." admitted that he had an insatiable need to make himself the center of attention. "If we consider all these processes together and the little tortures invented by these patients, one forms the impression that it is as if someone is constantly staring at them." Their "need to do forbidden and incongruous things" translated into the need to be seen, Rouart wrote of his patients.[21]

Above all, Rouart insisted, these children were "actors." They were simultaneously "spectators and censors." Their personalities often reflected similar characteristics found in their families. Stressing "the infantile structure of the tic itself" and "the infantile character of the tiqueur," Rouart argued that therapists treating an adult ticcer must return to the patient's childhood and identify "the other persons of the family drama" who had formed "the internal drama of the adult" ticcer.[22]

Because these connections often were not always obvious to the inexperienced observer, Rouart shared his analysis of seven-year-old Albert, as a case in point. On the surface, there were no signs of familial antecedents to Albert's tics, except that the mother was extremely nervous. The tics first appeared as throat clearings combined with shoulder shrugs when he was four or five. Two years later, when Rouart first saw him, Albert's tics were much more pronounced and varied. All of the boy's movements seemed to be reactions to his mother's punishments, including often having to sit in a chair without moving. While Albert outwardly showed passivity toward his mother's attempts to immobilize him, inwardly he was angry. The tics, therefore, revealed this tension, while at the same time, they attempted to provide a physical outlet for these imposed restraints. During play, Rouart said, Albert never ticced. For Rouart, Albert's story revealed "the role of the mother's opposition that had brought forth the tics," for the more she had attempted to immobilize Albert, the more "accentuated" his movements became.[23]

If overbearing mothers contributed to the formation of tics, so did familial disintegration. Rouart reported the case of a seven year-old boy who, as a result of his parent's separation, was raised by his grandmother and rarely saw his mother. The child developed tics, according to Rouart, because the grandmother had feminized the boy. Evidence for this was tied to the child's preference for "girls' games," especially playing with dolls and "going to sleep with a doll in his arms." When the boy was three or four years old, his grandmother chastised him for masturbating, "threatening to take him to the hospital where they would cut it [his penis] off." Following this threat the boy developed a series of problems, including defecating in his pants combined with a continuous and powerful sense of shame. Two weeks after this event, his tics appeared. Rouart's treatment focused on separating the boy from his grandmother and having the mother assume child-rearing duties, while the father reunited with the family. This approach, according to Rouart, was effective. The boy's school performance improved and his tics almost completely disappeared.[24]

In Rouart's estimation, this boy was "typical of the character of a child tiqueur." Although the case was "less pure than the preceding cases in terms of psychogenesis, since we found hereditary factors, obstetrical trauma, and a slight motor deficit," it demonstrated even more clearly the importance of psychological factors. This case confirmed, concluded Rouart, the essential outline of the psychoanalytic theory of tics—"the substitution of the tic to satisfy narcissistic repression." Moreover, "the powerful erotic and anal fixation manifesting itself by a compulsion to defecate in the pants [was] in contrast" with the boy's manifestly "meticulous cleanliness" and with his "possessive tendencies." Finally, Rouart noted, was that "as is the rule," the tics revealed the boy's repressed "aggressiveness."[25]

This was the only article that Rouart ever produced on tics,[26] but his arguments would be referenced, with the reverence often afforded a pioneer, by those French psychoanalysts who followed. Looking back twenty-five years later, Marseilles psychiatrists Joseph Alliez and Serge Audon credited Rouart's 1947 publication with creating a renewed interest in "la maladie des tics" and with reformulating a psychoanalytic understanding of tics in general.[27] Jean Delay never wrote again about tics. Although he later would attain international recognition for his work on the psychopharmacology of schizophrenia, his wartime work on tics was ignored in his own country and rarely cited even by those in other countries who argued for a neurological substrate for ticcing behaviors. The fact that the Nazis and their Vichy allies had attempted to eradicate psychoanalytic theory in favor of a reconstituted Nazi psychology informed by eugenic explanations would influence (in a negative way) the French psychiatric community's attitude toward most organic explanations of tic behaviors for the next half century.[28] To this day psychoanalysis remains strong in France because it portrays itself as a bulwark of defense against detested racist medical practices. Criticisms of psychoanalysis are often labeled as proto-fascist and anti-Semitic.[29]

Postwar Views

Rouart's influence was evident even in those studies that seemed to suggest organic and behavioral causes of tics. Girard and Schott's 1948 report (discussed earlier), claiming that they had "cured" two left-handers afflicted with motor and vocal tics by a "left-hand reeducation," emphasized that their finding in no way contradicted the claims of Rouart.[30] Instead, they insisted that their conclusions supported Rouart's because the latter had shown *"the importance of the familial milieu* in the genesis of tics," which is "perfectly reconcilable with our thesis."[31] Likewise, in another article entitled "Les Tics," Girard again suggested a physiological intervention while simultaneously endorsing Rouart's psychogenic claims. In fact, Girard pointed out, "very often the only notable element [in the afflicted] is the existence of psychopathology among their parents," many of whom "very frequently are ticcers themselves."[32]

In 1951, Serge Lebovici, who over the next four decades would emerge as the leading French child psychoanalyst, published a short but extremely influential book, *Les Tics chez l'Enfant*, which reverentially drew on Rouart's analysis.[33] Lebovici, a French-born psychoanalyst whose Jewish-Roumanian émigré father perished in a Nazi extermination camp, had clandestinely and illegally continued to practice his profession during the occupation of Paris. After the liberation, Lebovici emerged as a leader of the Société Psychanalytique de Paris, the French branch of the International Psychoanalytic Association.[34]

Les Tics chez l'Enfant did not present any clinical cases. Rather, it retraced the contested definitions and claims concerning the etiology, diagnosis, and treatment of convulsive tics from the nineteenth century until 1951. Lebovici provided, by way of textual exegesis, a persuasive, logical, and enduring explanation for clinicians treating children with motor and vocal tic symptoms. The result was a codification of the French psychoanalytic view of tics that continues to influence and inform French psychiatry to the present time.

Les Tics chez l'Enfant begins with a history that contextualized Gilles de la Tourette as one of a number of observers who, following Trousseau, attempted to make sense of these seemingly spontaneous movements and vocalizations. Drawing on Rouart, Lebovici emphasized Janet's view that "tics among children [were] the 'signal symptom' of an obsessional psychoneurosis."[35] Like Rouart, Lebovici found French psychoanalysis (Freud tempered by Janet) more insightful than its transplanted American cousin.

In contrast to Mahler, Lebovici wrote, "Brissaud and his students [Meige and Feindel] had attached . . . a greater importance to the relationship between tics and obsessions." Lebovici believed that childhood tics were most often motor expressions of obsessional rituals, but the tics often constituted only a portion of a ritual behavior, which might include other actions that did not fit the category of tics. An exclusive focus on tics might result in misreading the wider psychological issue in which the tics played a partial and symbolic role. Lebovici also rejected Mahler's separation of maladie des tics from those tics that she called "symptom tics." "In maladie des tics, according to the American psychoanalysts," explained Lebovici, "one has to deal with one of these organ neuroses; the tics are no longer the motor expression which symbolizes these conflicts, but the physiological expression of emotional constellations. The child ticcer has, according to the psychoanalytic terminology, a sick 'ego.' The tics express these troubles at the level of psycho-motor integration [and] to what they [these authors] call the damming-up of kinetic needs."[36]

Lebovici preferred Rouart's and Janet's view that a tic can best be understood as a compulsive manifestation. That is, the tic has both "a symbolic and substitutive value." This meshed with Freud's belief that "compulsive acts are a compromise." Even so, Lebovici conceded that these particular psychoanalytic explanations did not exhaust the possible interpretations of maladie des tics. Beyond the symbolic content, it was always possible that these behaviors were psychosomatic, reflecting "global psycho-motor troubles." Lebovici conceded the importance of studying the "physio-pathology or neuro-dynamic of tics." Yet Lebovici added that any "psychopathological study of tics was necessarily theoretical" because there was no evidence of anatomical abnormalities in children afflicted with tics.[37]

Asserting this lack of "anatomical documentation," Lebovici's definition of what constituted a psychosomatic explanation of tics, like Mahler's, ultimately privileged psychotherapeutic over somatic interventions. Returning to Mahler, Lebovici summed up and rejected her theory "that tiqueurs are (organically) predisposed," but that tics were activated in those predisposed who had experienced subsequent psychosexual conflicts. Instead, citing Gilles de la Tourette's critics, Lebovici argued for two types of somatic substrates for tics. The first were "transient tics" that arose in those with congenital nervous habits, as laid out in the work of Brissaud. Like Brissaud, Lebovici believed that these tics often could be cured by education. The second type were "chronic tics," which appeared in those with obsessional illnesses and, to a lesser extent, in those with conversion hysteria. If the obsessive type reflected the observations of Guinon, Meige and Feindel, and Janet, conversion hysteria, as it related to tics, had been laid out by Freud in his discussion of Frau Emmy von N.[38]

Most important for those who would be influenced by this work was that by Lebovici's criteria there was no place for a disorder in which the *tic itself* was the pathological feature of some [even idiopathic] organic condition. Rather, tics were always a symptom or sign of a bad habit or of a more serious psychopathology. From such a perspective there was no place for the separate tic illness described by Gilles de la Tourette. As a result, Lebovici's followers rejected all attempts to resurrect Gilles de la Tourette's disease because, by definition, it challenged Lebovici's reading that tics arose either in those with infantile habits or among those with obsessional or hysterical diagnoses. Building on Rouart, Lebovici had eliminated once and for all Gilles de la Tourette's typology. This helps to explain why those who later attempted to resurrect it on French soil, even as a syndrome, would face several levels of resistance.

As for Lebovici's view of treatment, transient tics were most amenable to the motor reeducation methods described by Brissaud and his colleagues. Chronic tics were more problematic. Even though these tics had "a certain spontaneous tendency to remission, the most complicated therapeutic efforts are justified," according to Lebovici, because "we have seen that this remission was not at all in itself meaningful to the cure of the obsessional neurosis." Lebovici endorsed Janet's view of "the favorable influence of calmness on tics," that "happiness was one of the best therapeutic additives for tiqueurs. Conversely, worry, strong emotions, and, generally, serious emotional shocks, even minimal ones," could cause "the reappearance of tics that have disappeared or the aggravation of those which exist." Lebovici urged that "the causes of excitation should be eliminated or reduced." Activities such as "reading and exciting games should be avoided." Because "bed rest can be in-dispensable, it is generally necessary to institute isolation in a spe-cial facility. This isolation," wrote Lebovici, should result in "a rupture of all contact with the family which must accept the sup-pression of visits, letters, and of packages."[39]

Like Mahler, Lebovici was certain that neurotic mothers were the vehicle of transmission of ticcing behaviors to their children.[40] It was essential "to avoid all those difficulties among these emo-tional and anxious children that could be provoked when their training is directed by their anxious mothers." Lebovici noted that all current experts agreed that mothers who were "excessive[ly] worried about cleanliness" and "created an atmosphere of exag-gerated 'perfectionism'" tended to restrict their children's locomo-tor development and thus generated the child's tics. "This knowl-edge," Lebovici urged, must be widely shared, so that "durable and difficult to treat" tic illness, which is in reality an obsessional disorder, can be understood.[41]

As his influence grew, Lebovici often would return to the subject of convulsive tics, renewing and revising the edges but not the core of his views.[42] Given Lebovici's enormous prestige, not only among French psychiatrists, but also with the very large number of French academics for whom psychoanalysis remains a central referent, *Les Tics chez l'Enfant* continued to be authoritative into the 1990s.

Throughout the 1950s the vast majority of French psychiatrists accepted Lebovici's paradigm that tics presented on a spectrum from transient to chronic forms and that parents', especially moth-ers', anxieties determined the tics' persistence.[43] Most important was Lebovici's insistence, as argued earlier by Janet and Rouart, that the chronic tics and involuntary vocalizations were symptoms of obsessive-compulsive neurosis. To demonstrate the phobic and obsessional nature of tics, three psychoanalysts, P. Geissmann, A. Lévy, and Lucien Israël, showed a film of a fifteen-year-old girl with florid motor tics and coprolalia to attendees at a psychiatric meeting held in Strasbourg in 1959. The trio believed that the pa-tient's eye blinking revealed that she was frightened, and thus, the "film demonstrated the intentionality of her abnormal move-ments." Her barking interspersed with her coprolalic exclamations "made it difficult to understand her," but revealed the regressive nature of her behaviors. The recurrence of the girl's symptoms was an expression of her "obsessional ideas and phobic impulses," confirming Lebovici's view that those with chronic tics fit a diagnosis of conversion hysteria.[44]

The young woman in the film was the subject of a thesis produced the next year by Israël's student, Simone Sylvie Bloch. Bloch identified the patient as "Marie-Thérèse B.," whose tics and coprolalia had first appeared when the girl was nine years old. The illness waxed and waned and the symptoms varied. At times Marie-Thérèse's motor tics were replaced by obsessions and compulsions. Ruling out any infectious or neurological etiology, Bloch emphasized "the reversibility of the symptoms, the balance between tics and obsessions, and the importance of the affective content of the symptoms, which turns us in favor of a theory of a psychogenic pathogenesis, without excluding, however, a certain nervous predisposition, perhaps a general dysfunction of the autonomic nervous system." Drawing on the work of Rouart and Lebovici, Bloch concluded that Marie-Thérèse's "case shows us that the patient presented a neurotic complex where the obsessional structure dominates (as numerous authors have remarked in analogous cases) with hysterical and perverted components." Thus, the girl's eye blinking was a masturbatory equivalent, while it simultaneously represented her guilt for "her desire to kill her mother, taking her (mother's) place with her father." In addition, the patient's tics, coprolalia, and general behavior were substitutes for her repressed aggression. Likewise, Marie-Thérèse's masochism and eroticism were actually "defense mechanisms" for these aggressive desires. The function of the tics was to release her internal motor tensions brought about by these psychogenic conflicts. Although Bloch believed that the patient's prognosis was "guarded," she was not persuaded that Marie-Thérèse necessarily faced a future of continued deterioration. Rather, as the patient grew to understand the symbolic meaning of her symptoms, she could look forward to "a reduction of her anguish and psychomotor problems."[45] Because no follow-up ever appeared, it is impossible to determine whether or not Marie-Thérèse's symptoms ameliorated.

Although French psychoanalysts seemed of one mind in their explanations of the etiological dynamics of tics, like their North American and British cousins, French practitioners' beliefs in the success of their interventions seemed to vary inversely with the length of their follow-ups—the shorter the follow-up, the greater the claim of therapeutic cure. French psychiatrists continued to modify and supplement their two main forms of intervention, reeducation and psychotherapy. Claims of cures by behavioral modifications seemed to wax and wane as frequently as the affliction itself. One such intervention, called "la cure de sommeil," first developed in the 1920s, consisted of awakening ticcing children from their sleep and subjecting them immediately to behavior modification. The assumption behind this procedure was that unconscious resistance was weakest during sleep and therefore behavioral (Pavlovian) interventions would have their greatest opportunity for success if they were instituted as close to the sleeping stage as possible.[46] As with la cure de sommeil, success of any form of behavioral modification was read as further evidence for the psychogenesis of tics.[47]

The Exceptions

There were two very important exceptions in the 1960s to the consensus among French researchers that convulsive tics ultimately should be understood and treated within a psychogenic frame. The first appeared in a 1965 article written by Toulon neurologists M. Langlois and L. Force, who argued that the symptoms described by Gilles de la Tourette resulted from a sequel to a rheumatic streptococcal infection, similar to the mechanism for Sydenham's chorea established in 1956 by Angelo Taranta and Gene H. Stollerman.[48] Because Langlois and Force were convinced that this process had resulted in lesions in the striate cortex, they also endorsed Seignot's findings about the effect of haloperidol in controlling tics. They treated a six-year-old patient with a combination of antibiotics, anti-inflammatories, and haloperidol[49] for control of her tics and coprolalia. Langlois and Force declared an immediate success in curing their patient, but, like so many other similar claims, they did not publish any follow-up of their patient's condition beyond the months immediately following her treatment.[50] Although the implications of their interventions seemed to invite a parallel critique of psychoanalytic assumptions, Langlois and Force did not use this occasion to directly criticize their psychoanalytic colleagues.

The second exception to the psychoanalytic paradigm was Yves Ranty's 1967 Bordeaux medical thesis, which concluded that Gilles de la Tourette's disease was an organic and not a psychogenic disorder. Ranty based his conclusions on extensive examination of four males, ranging from ages fifteen to sixty, all of whom had florid symptoms, including extensive body tics, coprolalia, copropraxia, and echolalia, that had waxed and waned for a number of years. As Ranty demonstrated, each case history could easily have been interpreted within a psychoanalytic frame. But, finding that each patient's tics could be controlled by neuroleptics, especially haloperidol, Ranty rejected a psychogenic etiology and instead insisted that the symptoms "must be the expression of a gross brain lesion." These "diverse abnormal movements," Ranty wrote, "have common characteristics in their strange gesticular pantomime that can be mistaken for hysterical or simulated movements." However, Ranty reminded his readers, similar symptoms have been observed among choreics, which we know are caused by "lesions of the caudate nucleus associated with encephalitic lesions of reticular formation of the thalamus, and perhaps of the hypothalamus."[51]

Building on the work of Taranta and Stollerman, which tied the malfunction of the basal ganglia to Sydenham's chorea, Ranty, like Langlois and Force, argued that the tendency to "differentiate varieties of abnormal movements, . . . in particular Gilles de la Tourette's disease," into separate categories was mistaken. Rather, he was persuaded that Gilles de la Tourette's disease "must be rejoined" with other involuntary movement disorders including choreas.[52] Ranty suggested that there were two general etiological routes to these lesions, acute infection and neurological degeneration.

Ranty pointed out that exacerbations of symptoms in two of the men could be tied to recurrent infections. He was convinced, as Langlois and Force had been two years earlier, that the symptoms of Gilles de la Tourette's disease also should be connected with an "acute infection, [as] during Sydenham's chorea, in which the agent would be streptococcus or staphylococcus, and in which inflammatory phenomena of the brain stem can regress." This explained the persistence of infantile motor movements later on. In contrast, "degenerative Gilles de la Tourette's disease" was similar to the course of Huntington's chorea in terms of its neurological progression, but without Huntington's hereditary history. Although Ranty admitted that his conclusion remained a "hypothesis," he was, nevertheless, "almost certain of the organic etiology of this affliction that today, if we cannot cure, at least we can stabilize and ameliorate in a spectacular fashion" with drugs.[53]

Ranty's work with his four patients led him to seek a specialty in psychiatry, also at Bordeaux Medical School, which required a second thesis. His thesis director, psychoanalytic psychiatrist J. M. Léger, was convinced that in his first thesis Ranty had not appreciated fully the psychogenic aspects of his patients' illnesses, and he persuaded Ranty to focus his psychiatry thesis on these issues. In 1969 Léger joined with his student to publish the preliminary results of Ranty's forthcoming psychiatry thesis. Contrary to Ranty's earlier thesis, he and Léger endorsed Bloch's study as "having convincingly developed the psychiatric aspects of tic in contrast with the neurological view." In line with this revised view, Ranty no longer advocated a connection between infection and tics. Reviewing Ranty's cases, Léger and Ranty wrote that in all four cases, "we have found a certain number of constant characteristics in all our observations: at bottom, almost always, there is a conflict with a domineering, over-protective, perfectionist, rigid, and at the same time anxious mother, and a subdued, spineless, weak, and dominated father. Often the patient was afraid of being excluded from parental love and manifested a certain sibling jealousy." Moreover, Léger and Ranty claimed that "the personality of these ticcers is almost always of the obsessional type described by Janet: some are hesitant, scrupulous, meticulous; others are obstinate, stubborn, have no confidence in themselves, they want to be listened to, and cannot stand to be contradicted." Most important, the authors, building on Rouart's view, emphasized that the ticcing and coprolalic symptoms, whatever their enabling organic cause, had become for each man "a language filled with signification. It expresses at the same time repressed masturbation, guilt, aggression, and masochism. The vocal tic expresses, through dirty words and impulsive phrases, a conflict at the Oedipal level where the parent of the opposite sex seems to be the central person."[54] In an English-language version of this paper that appeared several years later, Léger and Ranty expanded their claim that tics and other motor movements were a symbolic language. A tic, they insisted, was an integral part of a person's psychic life; "it is not a simple mechanism of execution, but each stage, each degree of organization, is the immediate expression of the connections that have settled between the human being and his surroundings."[55]

Reexamining the results of pharmacological treatment, Ranty and Léger (1969) discounted haloperidol's effect as evidence of an organic etiology of ticcing behaviors. As evidence, the authors cited their patients' unwillingness to tolerate haloperidol's often debilitating side effects, such as extreme lethargy, or what Léger and Ranty described as their patients' "attitude . . . toward their therapy." Although "haloperidol clearly diminished the frequency and intensity of tics," patients often "interrupted the therapy, on the pretext that the [drugs] were ineffective." For instance, one patient cited a depressive episode following treatment with haloperidol because the elimination of the tics resulted in an overwhelming sense of loss. "He abandoned the therapy, ticcing once again and was cured from his depression."[56]

By 1970 Ranty had come full circle. His psychiatric thesis, "L'Aspect Psychiatrique de la Maladie des Tics de Gilles de la Tourette," was a more detailed version of the 1969 essay coauthored with Léger. "The same problem," explained Ranty, "can be looked at in different ways." Ranty had chosen to approach Gilles de la Tourette's disease in his medical thesis "from a neurological angle," while in "this memoir" he looked at it "from a psychiatric angle." But, Ranty asserted, "it is not necessary to contrast these two aspects, but to the contrary to make them into a synthesis, in order to give the rare and curious affliction a holistic pathogenic interpretation."[57] However, that holism authorized those psychogenic views that Ranty's earlier dissertation had unambiguously rejected. In his 1970 thesis Ranty reread his earlier four cases in the context of French psychoanalytic interpreters, emphasizing how the organic features that had enabled the tics—the focus of his 1967 thesis— were always transformed in the service of psychogenic conflicts.[58] By 1970 Ranty believed that the therapeutic emphasis must be transferred from pharmacology to psychotherapy.[59] If not a total reversal of his earlier views, certainly Ranty's 1970 dissertation made him an acceptable interpreter of these symptoms within the psychoanalytic profession. Beyond this, Ranty's change of mind had revoked the only systematic French clinical study that had confirmed Langlois and Force's claims.

Keeping a Place for Psychoanalysis

That same year, Parisian pediatric psychiatrist Marie Olivennes reported a case of Gilles de la Tourette's disease in a twelve-year-old boy, Jean-Pierre, who had suffered a birth injury that had resulted in several convulsions during the first few days of life. Tics and behavioral problems appeared at age eight. These were combined with obsessions and compulsions that were so severe that for a time institutionalization was considered. However, Olivennes decided after interviewing the boy's parents that although the symptoms seemed to have an organic basis, they were actually the result of a psychosis brought on by an overprotective mother. Olivennes also prescribed chlorpromazine for Jean-Pierre, but she was convinced that it was the psychotherapy that led to improvement of tics. The drug, she asserted, only helped to manage the psychotherapy.[60]

By the 1970s many French psychiatrists were as conversant with the action of haloperidol and other antipsychotics as their North American counterparts. But most, similar to the American psychiatrists, were reluctant to concede that the ability of these agents to control tics and coprolalia was conclusive or even persuasive evidence that Gilles de la Tourette's disease should be treated as an organic disorder. Even those who conceded an underlying organic condition as enabling these symptoms, nevertheless continued to advocate that these organic substrates were only transformed into tics and coprolalia by psychogenic conflicts, which required psychoanalytic interventions.

In 1970 Vauclaire psychiatrists M. Yvonneau and P. Bezard reported the case of a twenty-three-year-old woman whose multiple motor tics, barking, and coprolalia—she cursed in Latin!—"were spectacularly ameliorated by [the dopamine antagonist] sulpiride (Dogmatil)." But Yvonneau and Bezard were convinced that the drug had no impact on the woman's "psychoneurosis," which they saw as the underlying cause of the behaviors. Reviewing the effects of sulpiride and haloperidol, Yvonneau and Bezard wrote that "these medications seemed to have an effect of only suspending Gilles de la Tourette's disease." This "pharmacological cure can be intermittent or even of a short duration in less severe tics." This is because an "organic substrate for Gilles de la Tourette's disease remains uncertain, and, if a new drug substance (Dogmatil) seems, at a high dosage, to suspend the clinical expression of this disease, it seems to be left to psychotherapeutics to take care of unraveling the prepsychotic origins of these illnesses." Discounting the woman's childhood convulsions and recurrent tonsillitis, the authors emphasized that the psychodynamics of this case revealed a conflicted family structure due to a weak and dominated father.

The patient's symptoms were "a narcissistic protest against this injustice" and were authorized by the patient's connivance with her mother.[61] For a parallel case, Yvonneau and Bezard cited the 1968 report by American psychoanalysts Hilde Bruch and Lawrence C. Thum connecting the severe tics and coprolalia of a twelve-year-old boy to a "maternal psychosis."[62]

In 1972 Joseph Alliez and Serge Audon published a comprehensive overview entitled "Reflexions sur la Maladie des Tics de Gilles de la Tourette." They rejected the view of Langlois and Force that these symptoms should be seen as an alternative manifestation of acute chorea: Gilles de la Tourette's disease, they insisted, was not a form of chorea because, unlike choreas, there was no evidence of a brain lesion of postinfectious origin. Alliez and Audon were not persuaded that the recent results achieved with pharmacology proved that the disorder could be reduced to an organic etiology. As evidence, they pointed out that "Ranty, who in 1967, had attempted to illustrate the organic thesis," had two years later, when he "examined tic illness from the perspective of its self-significa tion," concluded that "psychomotricity was a language revealing the particular aspects of the personality of the ticcers." This perspective revealed the hysteria, perverse exhibitionism, sadomasochism, frustration, and characteristic parental conflicts of the afflicted and exposed the compulsive nature of the movements. Ranty's findings reminded Alliez and Audon of "the frequency of obsessional symptoms of the type that Janet had earlier classified as symptoms of 'psychoasthéniques': hesitation, habitual doubt, lack of self-confidence, timidity, excessive scrupulousness, [and] superstitious tendencies." This group of symptoms, the authors believed, made it essential to focus on *the family milieu* of the ticcer."[63]

However, like Yvonneau and Bezard, whom they cited, Alliez and Audon could not ignore the "remarkable" effect of psychotropic drugs on ticcing symptoms. They concluded that the etiology of Gilles de la Tourette's disease was a combination of organic and psychological factors, "clearly, but inseparably linked." It was "an expression at once motoric and psychic: an anachronism, something a little unusual in our nosology."[64]

Evidence continued to mount, particularly from North American researchers, that haloperidol was more effective than psychoanalytic interventions. Reviewing the newest work, particularly the numerous studies by Arthur and Elaine Shapiro, Alliez and Audon published a second historical and clinical review in 1976 in *Annales Médico-Psychologiques*. Conceding that there was no longer any doubt that neurotransmission played a role in the presentation of these symptoms, Alliez and Audon nevertheless found that explanations of what that role was remained fragmentary and contradictory, requiring more research. Certainly, they were convinced that the effect of haloperidol did not rule out psychogenic and social factors. Alliez and Audon remained persuaded that convulsive tic disease should be understood in "a multidimensional framework where one considers neurological, psychiatric, psychological, sociological, psychoanalytic, and cultural approaches, in order to better exploit the dynamic aspect of all these factors in a therapeutic sense, without always juxtaposing them."[65]

By the 1980s, French psychoanalysis, split internally between orthodox and Lacanian factions and threatened externally by the psychopharmacological revolution in treatment, regrouped around the issue of convulsive tics.[66] Increasingly feeling on the defensive, the official psychoanalytic position toward an integrated view of tics and coprolalia became less flexible and more orthodox, privileging psychogenic factors. By the 1990s, leading French psychoanalysts would take the offensive, decrying organic explanations of Gilles de la Tourette's syndrome as a construction, almost a conspiracy, of the North American medical establishment.

The first shot was fired, appropriately enough by Serge Lebovici, at the January 1982 meeting of the French Society of Psychiatry of Children and Adolescents in Paris, which was devoted to "les tics chez l'enfant." Topics included children's tics as possible psychopathological symptoms, transitory behaviors, obsessive-compulsive behavior, nervous habits, and the changing psychopathological structures underlying tics. Also discussed was whether tics should be classified as psychomotor dysfunctions or psychopathological neuroses. Central to the discussion was the role of psychoanalysis and psychotherapy in the face of the success of chemotherapy for the treatment of tics. According to Lebovici, part of the conflict over psychoanalysis versus haloperidol could be tied to classification of symptoms. "It is difficult," he believed, "to ascertain the limits between tic, properly speaking, and nervous habits that can become veritable obsessions." It was important to recognize that "the psychopathological structure that underlies tic is

differently valued and it is variable. Patients with obsessional symptoms tend to be emphasized. But one can observe tics in obsessional
neurotics or in grave phobic syndromes, as in certain psychoses."[67]

Complicating the issue were trends coming from North America.
"Today," asserted Lebovici, "we must measure the respective effects of psychoanalytic cures and psychotherapy, on the one hand,
with chemotherapy with haloperidol on the other hand. In the
United States, and we will see this in France, growing importance
is given to the latter [haloperidol]," often "with excessive therapeutic risks," because Americans will not tolerate "anything that
interferes with the school and social life of the child ticcer." Lebovici suggested that this was the result of "the direct participation
of families at the therapeutic level." This interference with basic
clinical research "has appeared as a new phenomenon, most represented in Anglo-Saxon countries." Clearly, this was not a trend
that Lebovici favored. As he noted sarcastically, "in the United
States, the *'Tourette Syndrome Association'* publishes a *'newsletter'*
where summaries of research and publications of abstracts of social
meetings are destined to be picked over by the '100,000' Americans
who suffer from la maladie des tics."[68] Lebovici assumed that the
interference of laypersons with diagnosis, treatment, and research
was so outrageous that his colleagues needed only to learn of it in
order to be appalled at this trend in American medicine.

Lebovici's paper signaled a new strategy for the defense of psychoanalytic theories of tic. The threat would no longer be portrayed
as psychopharmacological, but rather, fitting the more general
French national attitude of the eighties, the issue was transformed
to a wider defense of French cultural institutions from American
medical and cultural imperialism. If the American Tourette Syndrome Association provided the appropriate symbolic target for
the deficiencies of American popular culture, the American Psychiatric Association's *Diagnostic and Statistical Manual of Mental
Disorders* (DSM) classification system represented the small-
minded and literal view that American psychiatrists had of mental
processes. Following Lebovici's presentation, psychiatrist P. Moran
of Toulouse suggested that the Americans had inflated the incidence
of Tourette syndrome by including many conditions that in France
would not be classified as maladie des tics.[69] In addition, Martine
Lefevre joined with Lebovici to attack the DSM-III's alleged de-
scriptive precision. The DSM, they argued, had transformed disor-

ders like Tourette's into a behavioral scale in which psychogenic perspectives, necessarily based on a more general view of the mechanisms of mental pathology, were "totally overwhelmed."[70]

Lebovici's accusation that Tourette syndrome was a creation of the American Tourette Syndrome Association would be repeated by many prominent French psychoanalysts in the next decade including Lucien Israël and Cyrille Koupernik. From one point of view this accusation is essentially accurate. In France, lay patient and family support and advocacy groups remained almost nonexistent at least until the late 1990s, but in North America they were legion, drawing no doubt on a long tradition of voluntary associations reaching back to the nineteenth century. In the case of Tourette's, where practitioners increasingly blamed parents for their children's illness, beginning in the 1970s the parents themselves, convinced of the organic nature of their children's affliction, led a revolt against psychoanalytic assertions.

The Triumph of
the Organic Narrative

$\cdots$ | 10 | $\cdots$

In 1975 a worried mother thumbed through the October 1975 issue of *Today's Health,* a magazine published by the American Medical Association for its member physicians to place in their waiting rooms. Her nine-year-old son, Tommy (a pseudonym), was being treated by a Freudian psychiatrist. Six years previously, when Tommy had continuous eye blinking, his doctor had assured the parents that "nothing was wrong," that Tommy merely had acquired a "habit." The blinking spontaneously disappeared, but a few years later Tommy displayed "a tiny" eye and lip twitch, "a nervous cough," and "a shrug of the shoulder." When these too went away, Tommy's mother, Claire Gold (also a pseudonym), assumed these also were habits similar to his earlier eye blinking. Tommy was an excessively shy child; soon after his mother forced him to attend a schoolmate's birthday party, he "started to go through a grotesque pattern of movements. He jerked his right arm and leg out at awkward angles to his body." When ordered to stop, he "kept right on." Claire was "alarmed and then angry. There's obviously nothing wrong with him," she remembered thinking. "He's just getting back at me, in his childish way, for making him go to the party." The situation escalated, with Claire demanding that her son control himself and Tommy insisting he couldn't. Although she had never spanked her son, Claire Gold lost control and slapped Tommy. The gyrations continued. A call to Tommy's pediatrician led to an examination by a neurologist, who concluded that "the boy's symptom is most likely an emotional reaction to stress or anxiety." He told Claire that she shouldn't "nag and discipline" Tommy "so much about making friends." Tommy's father, Jack (also a pseudonym), endorsed the neurologist's suggestion, increasing Claire's guilt about having initiated Tommy's "habits." "Was I actually the cause of my child's nervous tics?" Claire wondered.[1]

However, within two weeks, "Tommy the Tic," as other children now referred to him, "began making noises—grunts, whistles, throat-clearings." When these increased, Tommy's pediatrician sent him to a psychiatrist with whom Tommy met "three times a week to talk about his 'emotional problems.'" But Tommy seemed to get worse. He developed what the psychiatrist called a "patterned tic." "Every few minutes he would lick his lips, wipe his mouth on his sleeve, then his forehead with his hand." He would also repeat the same word several times. By now neighbors kept their children from playing with Tommy; Claire and Jack's marriage was becoming severely strained. "Jack and I had no sex life and," Claire recalled, "absolutely no social life. We couldn't have company because Tommy's actions were so bizarre. And, we couldn't ask a babysitter to stay with him while we went out. And, we couldn't take him out with us because we never knew when he would start to shriek or shout." Also, the ninety dollars per week for therapy was depleting the young couple's savings.[2]

As she absent-mindedly stared at the front cover of *Today's Health,* Claire saw the headline, "A Nightmarish Disease: Tourette's Syndrome—Literally—Can Make a Person Scream." She read the story of an eighteen-year-old boy, Orrin Palmer, whose symptoms and case history were similar to Tommy's. The condition, called Tourette syndrome, wrote the article's author Sally Wendkos Olds, "is physiological rather than psychological in origin . . . Many researchers feel that a chemical imbalance in the brain, most probably an excess of dopamine, may be at the root of Tourette's syndrome." Orrin, reported Olds, was under the care of Dr. Arthur K. Shapiro of Cornell University Medical College in New York City. With his wife, psychologist Elaine Shapiro, Olds continued, he had successfully treated 250 patients with "haloperidol, a drug that often eliminates Tourette's syndrome." That was because haloperidol [Haldol] was "known to block the effects of the brain chemical, dopamine . . . in the basal ganglia, an area associated with coordination of movement."[3]

Two decades later Claire would remember how she "sat there stunned" and then "burst into the doctor's office waving the magazine. 'Look at this,' I shouted. 'It describes all Tommy's symptoms . . . and it says they come from a disease which can be treated with medicine! Why didn't you know this? Why didn't you tell me?'" Unfazed, Tommy's psychiatrist replied, "your son may indeed have Tourette's Syndrome, I considered it but even if that's the case, I still think it is the result of an emotional disturbance. I think therapy is the answer rather than pills."[4]

The article in *Today's Health* also mentioned the Tourette Syndrome Association that had been founded in 1972 and was located in Bayside, New York. That afternoon, Claire phoned the Association's office and within a few weeks Tommy was examined by Dr. Shapiro, who "had no doubt that my son was suffering from Tourette's Syndrome." And, Claire reported, "after a week on haloperidol, most of Tommy's symptoms disappeared."[5]

Claire Gold's story, which was published in *Good Housekeeping* magazine in September 1976, is emblematic, one might even say formulaic, of the numerous newspaper and magazine articles that appeared in the American press from 1971 to 1985. All related a similar story of children with bizarre movements and vocalizations, shunted from pediatrician to psychiatrist, whose parents were told that the behaviors were a result of their parenting. Then, after years of failed psychotherapeutic and behavioral modification treatments, each family learned, as if through some deus ex machina, the truth that these odd behaviors were caused by a chemical imbalance or neurotransmission malfunction, which was most likely amenable to treatment with haloperidol. Each became a patient in Dr. Shapiro's care.

In June 1972, the *Wall Street Journal* ran a front page story about Sheila, who at age twelve was called "Shellshock," "Twitchy," and "Shuckle-Shuckle" by her classmates. "Her tongue rolled, and her face, arms and legs twitched uncontrollably. From her mouth came dog-like barks and, worst of all, streams of muttered obscenities." After a series of misdiagnoses, Sheila saw Shapiro, who placed her on haloperidol. "Now 29 years old, Sheila is happily married and will soon be a mother; her bizarre symptoms have been all but erased for the past three years." Another "person whose life has been changed by the drug" was Donald, a New Jersey leather worker, whose symptoms began at the age of 13. His parents were told that "it was just a 'nervous habit' that would pass. When instead the symptoms got progressively worse, various doctors tried all types of 'cures,' . . . including electric-shock therapy. By the time he had reached his mid-40s, things had gotten so bad that doctors recommended a lobotomy . . . to stop his incessant twitching and cursing. Fortunately the man was referred instead to Dr. Arthur K. Shapiro . . . That was 1967, and Donald says that the drug has kept him '95% symptom free' ever since."[6]

The *New York Daily News* in December 1972 told about Stella, who "for years . . . had been afflicted with uncontrollable tics and horribly embarrassing habits. She would find herself sticking her tongue out, barking like a dog or grunting in the middle of a conversation. Worst of all was the obscenity. Words she could hardly believe she knew would drop from her lips at any time, in spite of herself." Like Tommy, Sheila, and Donald, Stella's family had "taken her to an endless round of doctors for treatment with an uncountable number of drugs and with psychotherapy. Nothing had worked. At the age of twenty-three, almost unemployable, Stella was coming to the end of her tether." Then, as with the others, "chance brought Stella to Dr. Arthur K. Shapiro—and a new life." Shapiro prescribed "a powerful new drug, haloperidol. Almost miraculously, her symptoms began to vanish."[7]

The narrative was retold by Jane Brody in a *New York Times* article in May 1975, this time focusing on a fifty-four-year-old ticcing patient, Robert P. "Like so many other victims of Tourette's syndrome," Brody reported, "before Haldol, Robert had tried everything—'psychiatrists, neurologists, hypnotists, chiropractors,' and he had even been considered a candidate for frontal lobotomy—all to no avail." The Shapiros were quoted as reporting "that it is not unusual for a victim of the illness to have been to 100 different doctors and to have spent up to $100,000 on ineffective therapy."[8] Brody's syndicated story was republished throughout the United States.[9] Similar testimonials appeared in the *Chicago Tribune* (1973), *Reader's Digest* (1973), and the *Philadelphia Inquirer* (1976).[10] Smaller circulation newspapers and Sunday magazine supplements published analogous reports.[11]

Although the general veracity of these narratives is indisputable, their resemblance in structure, form, and message was hardly coincidental. Often, they could be traced to a single source. On the page following the *Good Housekeeping* story about Tommy there was a quarter-page box insert containing the name and address of the Tourette Syndrome Association, explaining that "its members include individuals with the disorder, their families, medical authorities in the field, and others who are concerned. It raises funds to support medical research. It publishes information to help undiagnosed and misdiagnosed patients know the true facts about their disease. Through meetings and newsletters, members learn about new findings in the field and help each other bear the burden by sharing experiences." Readers were advised to write to the Association "for further information—including names of physicians in the U.S. who are familiar with diagnosis and treatment of the disease." In fact, the Association had cajoled and finally recruited the editors of *Good Housekeeping* to publish this piece, as it had convinced Sally Olds to write the story about Orrin Palmer in *Today's Health.*[12] By the time Claire Gold's story appeared, the Association had been successful in getting similar stories published in the national press. Each of the articles mentioned the Tourette Syndrome Association, including its address for those desiring information about the disease or a list of physicians who were competent to treat it.

Until the mid-1970s the physician the Association most often suggested for a referral was Arthur K. Shapiro, the psychiatrist almost always mentioned in these narratives. This was no conspiracy; rather, it reflected a history of mutual cooperation and interdependence that would prove crucial to the way Tourette syndrome would be understood and treated in North America and eventually much of the rest of the world. The Tourette narrative played a central role in communicating this new definition of the cause and treatment of multiple motor tics and involuntary vocalizations. The construction, dissemination, and power of the narrative had two equal and interconnected sources, the Tourette Syndrome Association and Arthur and Elaine Shapiro.

Toward a New Master Narrative

In April 1965 a twenty-four-year-old woman was referred for psychotherapeutic treatment to a New York psychiatrist, Dr. Arthur K. Shapiro. Shapiro, then Associate Clinical Professor at the New

York Hospital—Cornell Medical Center, reported that the woman's "symptoms were striking and bizarre: spasmodic jerking of the head, neck, shoulders, arms and torso; various facial grimaces; odd barking and grunting sounds; frequent throat clearing; and periodic and forceful protrusion of the tongue."[13] Although she displayed no coprolalia during the initial visit, the woman reported that at other times she could not restrain herself from occasionally shouting out the word "cocksucker." The woman reported that when she was ten, some time after she had begun biweekly injections of an unnamed medication for hay fever, her arms began to twitch, "shortly followed by twitching and jerking of the head and neck." Her symptoms increased and by high school her movements had led her classmates to refer to her as "Twitchey." Although they waxed and waned over the years, the twitching and jerking always returned, "insidiously worsened." At age twenty-two, "short, loud screams began," followed six months later by "throat-clearing, rasping and barking noises, and coprolalia." Although the woman managed to confine her screaming and cursing to when she was home, when out in public "other symptoms increased, such as throat-clearing, barking noises, and tongue-protrusion." Social tensions, wrote the Shapiros, exacerbated her symptoms. "A representative sample of symptoms occurring during one minute would include twenty jerks of the head, neck, and torso, two tongue-protrusions, and two grunts or barks."[14]

Both her mother and the patient attributed these behaviors to psychological factors. Noticing that the patient could control her symptoms for brief periods of time, the mother assumed that her daughter's problem was definitely not organic, a view shared by the referring psychiatrist, whose diagnosis was "habit tic with hysterical personality." The patient herself admitted that when she "occasionally saw someone twitching and jerking, her immediate reaction was that the person was crazy."[15]

Shapiro was convinced before he ever met a Tourette's patient that most psychiatric disorders that were treated as psychogenic were in reality organic. "We were always perplexed," remembered Elaine Shapiro, "by the contradiction between our absolute belief that this is neurological" and the seemingly psychological symptoms that the patients displayed. Elaine Shapiro[16] recalled that in 1965 when "this first patient" walked into his office, "I know that Arthur was absolutely convinced, he didn't have a shred of doubt that psychology played practically no role."[17]

Hospitalizing the woman for the next several months, Shapiro administered thirty-six neuroleptic and antidepressant drugs and combinations of drugs,[18] settling finally on the major neuroleptic tranquilizer haloperidol "because marked improvement had been reported [in medical journals] in five patients." For the succeeding eleven months he treated the young woman with dosages ranging from 1.5 to 10 milligrams per day. "From the first day of treatment," wrote Shapiro, the "symptoms disappeared." At the higher dosages there were, however, severe Parkinson-like side effects.

Figure 10.1 Arthur K. Shapiro and Elaine S. Shapiro, citing their successful results with the drug haloperidol, insisted that Tourette syndrome was an organic disorder and that psychoanalytic interventions did more harm than good. Their success in treating Tourette's patients with haloperidol received wide media coverage and provided the impetus for the formation of the Tourette Syndrome Association.

Shapiro finally settled on 3 milligrams per day as the most effective treatment.[19]

In their 1968 article reporting successful treatment of the young woman with haloperidol, Arthur and Elaine Shapiro not only argued that the etiology of these symptoms was an "organic pathology of the central nervous system," but also, at Arthur's insistence, presented their findings as evidence of the therapeutic and intellectual paucity of psychoanalytic psychiatry. In retrospect, it is not surprising, given the dominance of psychoanalysts on the editorial boards of American psychiatric journals in the 1960s, that the Shapiros' article was rejected for publication by every major American psychiatric journal, finding a home only in 1968 in the *British Journal of Psychiatry*.

The Shapiros' article laid out a sort of "master narrative" for all the media stories and testimonials that would follow. This patient's symptoms began at age ten and for the next fourteen years the many physicians who treated her "communicated to the patient and her family that emotional illnesses caused the symptoms." Although psychotherapy did nothing to relieve the woman's symptoms, it did have "a harmful effect on her fantasies and character formation." The woman's earlier diagnosis of "personality disorder with inadequate and passive dependent character traits" was, the Shapiros told their readers, "based on psychiatric impression."[20] In contrast, the Shapiros, certain that the condition was organic, treated their patient with haloperidol.

The Shapiros' original narrative did two other things. First, it gave the disease a name. Second, it made certain that the name, Gilles de la Tourette syndrome, would be synonymous with an organic disorder. The Shapiros emphasized Gilles de la Tourette's 1885 description of symptoms, but downplayed his view that tics had a degenerative etiology. They established the label "Gilles de la Tourette's syndrome" as the descriptor of a neurological disorder, which by definition stood in opposition to psychoanalytic claims. From the Shapiros' perspective, psychoanalytic treatment most likely "contributed to the patient's shyness, inhibited aggression, passivity, and fantasies of insanity."[21] The Shapiros decried (in a sentence that they would repeat in several articles) "the fashion in medicine to attribute symptoms and diseases without demonstrable organic pathology to a psychological wastebasket diagnosis." And, with an obvious reference to Mahler's patients, the

Shapiros warned that psychoanalytic treatment of these symptoms "may result in iatrogenic [physician-induced] psychopathology. Physicians should be sensitive and cautious about the possible harm to patients of premature psychological diagnosis."[22] Every element of the media narratives that would follow, except for the referral role of the Tourette Syndrome Association, are found in the Shapiros' 1968 article.

The aim of all of the Shapiros' work in the late 1960s and early 1970s was to demonstrate that those diagnosed with Tourette syndrome suffered from an organic rather than psychogenic disorder.[23] The most compelling evidence for such a conclusion was that the ticcing and vocal symptoms displayed by these sufferers could, in the majority of cases, be effectively controlled by haloperidol.[24] Although today haloperidol's action is understood as reducing dopamine transmission in the substantia nigra to the basal ganglia circuit, when he first used it, Arthur Shapiro had no explanation for its probable pharmacologic action. In fact, as late as 1994 he still had no idea how haloperidol actually worked—except that it "poisons" the system.[25] Of course, by the mid-1970s Shapiro was aware of the consensus that haloperidol reduced transmission of dopamine and what that implied about the role of the basal ganglia in Tourette's.[26] Nevertheless, until the end of his life, Shapiro saw haloperidol as a gross antipsychotic agent whose action could tell us very little about the etiology of the symptoms.[27] The Shapiros' insistence on an organic etiology of Tourette's essentially rested on their clinical experience and on their profound mistrust of psychogenic theories and psychoanalytic therapeutics.

These views were underlined in a symposium panel on Gilles de la Tourette syndrome organized and chaired by Arthur Shapiro at the May 1968 meeting of the American Psychiatric Convention in Boston. The panel transcription, which was published in the September 1970 issue of the *New York State Journal of Medicine,* was supported by a grant-in-aid from McNeil Laboratories, the manufacturer of haloperidol, and by a donation from H. B. (Bill) Pearl, who would later play a crucial role in funding the Tourette Syndrome Association.[28] Shapiro introduced and moderated the panel, made a presentation, and summarized the conclusions.

Rather than elaborating a possible scenario of the pharmacologic or neurobiological action of haloperidol, as the other panelists had done, Shapiro launched into an attack on psychoanalytic therapeutics. "The tendency of physicians to attribute Tourette's syndrome prematurely to psychogenic factors," Shapiro argued, "contributes to an iatrogenic onus to patients already so incredibly burdened that it is a wonder to me and a credit to them that they can maintain their sanity."[29] Reviewing the discussion, Shapiro reported that "most participants agreed that haloperidol was the best and most predictable of all available treatments." Often it abolished all tics and had proved between 30 to 90 percent effective, he said. "All panelists," wrote Shapiro, "believed that the syndrome was caused by neuropsychiatric impairment of the central nervous system, although several believed that coprolalia could not be explained neurophysiologically."[30]

The Shapiros' subsequent publications and papers elaborated the positions laid out in the 1968 paper and Boston symposium. Thus, before the first meeting of what would become the Tourette Syndrome Association (TSA) in a room provided by Shapiro at New York Hospital, the central thesis of the Association's official point of view already had been laid out. The Shapiros had reversed the rhetorical burden of proof. From then on, based on their results with haloperidol, they would claim that the organic factors of Tourette syndrome were established. Those arguing for a psychogenic etiology would have to demonstrate their conclusion through the scientific method, even though the Shapiros conceded that the action of haloperidol could not meet this test.[31] In a 1973 paper entitled "Organic Factors in Gilles de la Tourette's Syndrome," the Shapiros and their New York Hospital colleagues, Henriette Wayne and John Clarkin, concluded that "although considerable indirect evidence supporting an organic aetiology has been presented, definitive evidence is not yet available." But most emphatically "our data, and our interpretations of the literature, do not support a psychological aetiology, and the burden of proof is on those claiming it."[32]

Arthur and Elaine Shapiro's book, *Gilles de la Tourette Syndrome,* coauthored with psychiatrist Ruth Bruun and neurologist Richard Sweet, and published in 1978, is an impressive and comprehensive overview of what was known about Tourette syndrome. It concluded with a series of patient testimonials, each of which read like all the other testimonial narratives of Tourette syndrome. "The Tourette Syndrome Association put us in touch with Drs. Elaine and Arthur Shapiro," wrote a mother who along with her son had been diagnosed with Tourette syndrome. "We visited them in New York in June 1974 and were both diagnosed as having Tourette's. I cannot begin to tell you what our visit to them meant to us. It has been such a relief to know what we have. For the first time Danny and I were able to freely talk about ourselves and even joke about it. I can't stress too much how important it is for a person with Tourette's to be correctly diagnosed."[33] A father reported that "it was a relief to hear the phrase, 'Gilles de la Tourette syndrome' applied to our son's illness, after 27 long frustrating years . . . of counseling, psychiatric evaluations and testing, a month in a mental institution and, worst of all, four sad, frightening years in a home for emotionally disturbed children." The father was relieved that he was able "to finally meet with an unusually competent psychiatrist who stated with assurance that the tic symptoms did not indicate that our son was psychotic, neurotic, or emotionally disturbed because of his family environment and parental inadequacy." This revelation occurred because "an article in the *Wall Street Journal* describing an unusual affliction caught my attention." The information in the article "brought us to Dr. Arthur Shapiro and the Tourette's Syndrome Association." The father had hope for his son's future, "hope that he will eventually be cured of his affliction through new scientific research being conducted now."[34]

A mother testified to the long, fruitless search for diagnosis and treatment for her daughter. The journey was the same as the others, with physicians assuring the mother time and again that her daughter would "outgrow" her tics and hooting vocalizations. But the "habits" got worse. When the child's teacher complained that the girl's noises were disturbing other students, her parents became increasingly desperate. "For three months, five days a week, we took our daughter to a psychologist," who pronounced the therapy a success when the sounds ceased. But the parents were "bitterly disappointed one month later when the tics returned and were even worse than before." The daughter considered herself "a personal failure." Finally, with "the help of our psychologist we located . . . Dr. Arthur Shapiro who put our daughter on Haldol immediately. Although she suffered from many of the drug's side effects, she had about an 80 percent reduction in her tics." Like the others, this mother was extremely grateful. "At last my child has been diagnosed. She has Tourette syndrome. Somehow a monster is less frightening when you call it by a name."[35]

That the name was foreign and loosely based on a forgotten

disease classification provided additional authenticity. Like the others, this mother affirmed that the label "Gilles de la Tourette syndrome" reinforced her faith that this was an organic disease. Betti Teltscher, the TSA president, vice president, and board member (1974 to 1981), remembered her great "relief" when she first read that Tourette's was a "physical disorder." Teltscher, whose son was under Shapiro's care, had been "absolutely . . . much more willing to accept the diagnosis" because "it was physical." She was certain that if, instead, she had been told that Tourette's was psychological, would she have "felt a lot less relief at having a diagnosis." Years of experience in the Association had convinced Teltscher that "once we could say that this [TS] is neurological, an increased number of people would come forward because the stigma would be lifted from their shoulders as would the guilt associated with a progressive psychodynamic illness."[36]

The Tourette Syndrome Association

The Association's roots appropriately enough can be traced to a letter that appeared in the *New York Post* in December 1971, written by a distraught father, Martin Levey of Flushing, New York. "We have a son who has a very rare neurological disorder called 'Gilles de la Tourette's syndrome,'" wrote Levey to the *Post's* editor.[37] The Leveys, according to a *Wall Street Journal* article published the following June, had spent more than $50,000 in "an eleven-year search for a doctor who could cure their son" Bill, who "began making slight noises at the age of six." Soon, "the noises became obscenities, shouted at the top of his lungs, and were accompanied by body tics." Mrs. Levey reported that "the really tragic part of it was that the psychiatrists we went to would all blame me for Bill's problems. Right in front of me, they would ask him, 'What has your mother been doing to you?'"[38]

The Leveys asked "parents of children who have this syndrome" to "please contact us through the *Post*."[39] Although the letter appeared to be a desperate plea for medical advice, Bill had been under the Shapiros' care since 1967 when he was seventeen years old. By 1971, when Levey's letter was published, Bill had been taking haloperidol for four years and his symptoms had almost totally abated.[40] The purpose of Levey's letter was to organize a group to lobby physicians, the public, and the government to support acceptance of Tourette's as an organic disorder.

In early 1972, as a result of the letter in the *Post,* a few families met at the Leveys' home.[41] The group agreed on the necessity of an activist organization if they were to find suitable treatments and ultimately a cure for this disorder. The Shapiros agreed to inform their other patients about the group's existence and to obtain a room at New York Hospital for future meetings. Also, an invitation to parents with children exhibiting ticcing symptoms was placed in the *Long Island Press.*[42] Augmented to twenty-five persons, the group, with the Shapiros in attendance, held its first official meeting.[43] The families decided to hold regular meetings in which they would "share information on the illness, . . . comfort one another," and "consider how they might reach others who were suffering from the same malady."[44] The group decided to pursue five interrelated strategies: legal status, publicity, recruitment, medical advice and support, and research funding.

The group applied for and received a charter from the State of New York as a tax-exempt, nonprofit organization called the "Gilles de la Tourette Syndrome Association, Inc." In order to recruit members from what they believed to be a wider community of afflicted and their families, the group published personal notices in the *New York Times* inviting those who displayed ticcing symptoms, or whose family members displayed ticcing symptoms, to contact the Association.[45]

Do you have multiple tics?

Involuntary muscular actions, verbal noises, facial tics, repeating actions, or obscene words, foot stamping, head and shoulder jerking, occurring in combination and changing from time to time. For additional information please write to The Gilles de la Tourette Syndrome Association, Inc. Box 3519, Grand Central Station, New York, New York, 10017.[46]

As a result of this and other publicity, 140 members were added to the Association by July 1974. Later that same year the first "Gilles de la Tourette Syndrome Association Newsletter" [hereafter, TSA newsletter] appeared, reporting on a variety of issues.[47]

The group sought medical advice and support from sympathetic physicians. Arthur Shapiro had agreed to act as an adviser. In these early years Arthur and Elaine Shapiro were the main connection the group had with the medical community. In fact, the Shapiros played such a central role in the formative years of the Association that it is impossible to separate their views on the etiology and treatment of Tourette's from those of the Association's. After 1972, as each of the Shapiros' new studies appeared or a paper was delivered at a professional meeting, it would be presented simultaneously in Association meetings and its conclusions would appear in the TSA newsletter. Often, the TSA would reprint and distribute these studies to its membership, to those inquiring about Tourette's, and to physicians requesting assistance and information.[48]

The Association's members, most of whom had children under the Shapiros' care, were extremely appreciative of the Shapiros' work. The first official TSA newsletter (Spring 1974) contained a "tribute . . . to Dr. Arthur K. Shapiro and the many dedicated medical men and women who are responsible for all the research done at the Special Studies Laboratories of New York Hospital" and to "Dr. Elaine Shapiro for all her efforts in writing the forthcoming publication, 'Questions and Answers.'" The Association also "offered their heartfelt thank you" for Arthur Shapiro's "splendid cooperation" for providing "the use of the room for our meetings, and for the endless questions so patiently answered by all of the staff."[49]

The group agreed to supply the Shapiros with subjects for their clinical study of two drugs that seemed to reduce dopamine transmission, alpha-methyl-paratyrosine (AMPT) and tetrabenazine. Three other physicians joined the Shapiros on this project—Bruce Bienenstock, Ruth Bruun, and Richard Sweet.[50] By October 1974 the Association had also recruited Dr. Arnold Friedhoff of New York University Medical School. One hundred and twenty-five people attended the fall 1974 meeting held at New York Hospital. Friedhoff talked to the group about his research on dopamine transmission and why some patients failed to respond to haloperidol. Aside from the Shapiros, a number of other physicians attended this meeting, including Rosewell Eldridge, acting chief of epidemiology of the National Institute of Neurological Disease and Stroke in Washington, D.C.; Edith Miller, assistant director of neurology at Brooklyn's Maimonides Hospital; and Richard Sweet. The Association was also in contact with Faruk S. Abuzzahab at the University of Minnesota, who was creating an international registry of patients with Tourette syndrome.[51] Instrumental in arranging the fall 1974 meeting was Sheldon Novick, a psychiatrist in Manhattan and a practitioner without a personal research agenda, who had contacted neurologists and psychiatrists working in related areas as well as "the small number of professionals who had actually published on Tourette's."[52] Teltscher recalled that Novick, who would become the first medical director of the Association, "played an influential role in getting the medical community interested. He wrote to everybody he could think of that had the slightest connection in their research or papers with Tourette and he was responsible for encouraging a growing number of physicians to investigate the illness—including Oliver Sacks, [Rosewell] Eldridge, and Donald Cohen at Yale. So, I give him credit really, for the beginning of the Tourette Association."[53]

In 1972, the Association had applied to the National Institutes of Health for a grant, but was turned down because the reviewers believed that there were probably no more than one hundred cases of Tourette's in the entire nation.[54] By this time, few involved with the Association believed that Tourette's was a rare disorder, but no one had any idea of its actual prevalence. However, they all agreed that to obtain funding support, they had to alter the perception that Tourette's was a rare disorder. Certain from the overwhelming response to their advertisements and their personal experiences that the media and physicians had greatly underestimated the number of persons afflicted with Tourette syndrome, Association leaders made a determined effort to correct this misimpression. Judy Wertheim, another of the Association's founders and its president from 1980 to 1985, later confessed that she had "made up" the figure that Jane Brody published in her 1975 *New York Times* article that "conservatively, an estimated 10,000 Americans are victims." Wertheim was absolutely convinced that even this number was an underestimation.[55]

Spurred on by Novick, the group continued to encourage any opportunities for research. Most important was its ability to supply patients for researchers interested in studying disorders that might have some connection with Tourette's.[56] By January 1975, Novick had brought in Oliver Sacks, who had agreed to "begin . . . intensive clinical studies of Tourette patients and their families."[57] That same month, Melvin Van Woert, Professor of Internal Medicine at Mt. Sinai School of Medicine, talked to an Association meeting about the "impressive improvements" in muscle spasm patients he treated with serotonin and carbidopa, which in contrast to halo-peridol, increased the transmission of dopamine.[58] Novick also announced that Arthur and Elaine Shapiro had collaborated with Richard Sweet and Ruth Bruun to produce an article entitled "The Diagnosis, Etiology and Treatment of Gilles de la Tourette's Syndrome," which the Association had reprinted to distribute to patients and practitioners.[59] By the mid-1970s the Association had inextricably attached itself to and encouraged every new and promising route toward identifying and treating Tourette syndrome as an organic disorder. Notably absent were any projects that assumed that Tourette's had psychological causes or even studies that sought to treat the psychological problems that might result from the affliction.[60]

In April 1975 Bill and Eleanor Pearl donated office space for the Association in Bayside, Long Island. At the suggestion of Marjorie Guthrie, founder and president of the Committee to Combat Huntington's Disease, the cumbersome and difficult-to-pronounce "Gilles de la" was dropped, and the organization was renamed the Tourette Syndrome Association, Inc.[61] By April 1975, the Association newsletter listed and supported six ongoing and new research projects.[62] Simultaneously, it prepared exhibits at national and regional medical meetings. In May the Association sponsored a one-day workshop "for physicians only" at the Americana Hotel in Manhattan. Speakers included Abuzzahab, Bruun, Friedhoff, Arthur and Elaine Shapiro, and Henriette Wayne.[63] The Association had thus launched what would emerge as one of its central functions, providing venues for researchers to meet and exchange views about the causes and treatment of Tourette syndrome.

Meanwhile, the Association publicized Tourette's in an effort to recruit sufferers and to inform the public and the media that the affliction was an organic disorder whose many victims should be viewed with compassion. The general information mass mailer was entitled, "Tourette Syndrome: A Neurological Disorder Characterized by Multiple Tics."[64] A new one-page advertisement, "We're Looking for People Who Twitch, Yelp, and Grunt Uncontrollably," was created, and the Association persuaded a number of national magazines, including the *Saturday Review, U.S. News and World Report,* and *Medical Economics* to print it as a public-service announcement. The advertisement described typical symptoms, noting that "most victims accept it as an emotional problem, *which it is not*. Many doctors, too, misdiagnose it as a mental disorder. *But it is physical!* There is a drug that can help control the symptoms in many people." Readers of the ad were told that "if you or your children or friends show any of these symptoms, please contact us immediately for more information." Included was a tear-out return portion so that those who wanted more information or wished to be placed on a mailing list could send their names directly to the Tourette Syndrome Association office.[65] By July 1976, the *Saturday Review* advertisement alone had resulted in 250 responses.[66]

The Bayside office officially opened in June with a volunteer staff led by Eleanor Pearl.[67] The Association convinced the editors of the *Ladies Home Journal* to publish a brief piece on the syndrome in its April 1975 issue (which resulted in over one hundred letters requesting information from the Association).[68] By October 1975, the office had handled over three hundred requests for information from all over the United States and Canada, 90 percent of which were from families, the rest from physicians. In response, each was sent a copy of the Association's new brochure, plus copies of the Shapiros' 1968 article and the 1975 paper, which the Shapiros had coauthored with Sweet and Bruun.

Meanwhile, the publicity continued. Not only was the Association instrumental in providing information for full-length articles in national and regional magazines and journals, but also it was successful in placing smaller informational articles in the *Reader's Digest* (1973), *Science Digest* (1973), and *Psychology Today* (1976).[69] In addition, the Association inserted letters in the medical advice columns of *Newsday* and the *New York Daily News* whose answers publicized the group.[70] Eventually, this strategy resulted in letters published in nationally syndicated advice columns, including "Dr. Joyce Brothers," "Ann Landers," and "Dear Abby." The Association also got its message across through television and film. Arthur Shapiro appeared on the "Today Show" in June 1975.[71] That same year an Association board member convinced a New York University film professor to produce a thirty-second television commercial about the symptoms of Tourette's, which ended with the address and phone number of the TSA. McNeill Laboratories (the manufacturer of Haldol) agreed to provide funds for a teaching film for physicians. Along with each showing of the film, McNeill also agreed to distribute and pay for reprints of the Shapiros' and Wayne's 1973 article, "Treatment of Tourette Syndrome," as well as the Tourette Syndrome Association brochure.[72]

By now, what the Association considered favorable publicity began to appear on its own. In February 1976 the New York City affiliate of PBS aired a documentary on Tourette's as part of its program "51st State." The documentary emphasized the contributions of Arthur Shapiro, who was interviewed in the broadcast.[73] The Association increased its efforts to "educate" physicians about the organic nature of Tourette's by attempting to place exhibits at every relevant national and regional medical meeting. In addition, those active in Tourette's research, especially the Shapiros and Abuzzahab,[74] continued to make presentations about Tourette syndrome at national and regional medical meetings.[75] A medical advisory board was established, which in addition to Bruun, Friedhoff, Sacks, A. Shapiro, and Sweet, included Thomas Chase, director of scientific research of the National Institute of Neurological and Communicative Diseases and Stroke, and Robert Good, president and director of Sloan-Kettering Institute. By 1976 a new generation of medical researchers had been stimulated to examine organic etiological factors in Tourette syndrome. Among them were Malcolm Bowers and Donald Cohen of Yale Medical School; Gerald Brown of the National Institutes of Health; Harold Klawans, director of neurology at Reese Medical Center in Chicago; and Harvey Moldofsky, chief of clinical psychiatry at Clarke Institute (Toronto). The Association also obtained its first grants in 1976: $3,000 from the Schubert Foundation and $1,000 from the Danskin Foundation. In addition, eleven research centers were examining the relative effectiveness of a new dopamine antagonist, pimozide (Orap), for the control of tics.

By July, the copresidents, Teltscher and Bernard Klein, reported that an "education kit" had been created so that additional regional chapters of the TSA could be more easily developed. Already, chapters had been organized in northern and southern California, northern and southern Florida, Connecticut, Massachusetts, New Jersey, Illinois, Michigan, Maryland, and Washington, D.C. A separate Canadian affiliate, called the Tourette Syndrome Foundation, was also established. Letters poured in from other venues as far away as Johannesburg.[76] But, as the October 1976 TSA newsletter reported, "The most extraordinary publicity we have received to date has been the article which appeared in *Good Housekeeping Magazine,* September 1976. We expect to have answered 2,500 inquiries by mid-October. It is our feeling that the majority of people writing to us in response to this article have a Tourette patient (undiagnosed) in their family." By the end of 1976, each of the goals outlined by the group that had met at New York Hospital in 1972 was firmly established. If Tourette syndrome was not yet a household word, an extraordinary transformation in attitude of patients, families, and physicians was underway.

In 1979, through the efforts of board member Abbey Meyers, a mother of three children with Tourette's, the Association was awarded a grant by the New York State Department of Developmental Disabilities to set up a "model program" in New York State. The $30,414 annual grant (renewable for three years) allowed the Association to hire a "coordinator who will also act as a case-worker for TS patients who need assistance and advocacy in education, employment, housing and financial problems."[77]

By 1982, in operation for a decade, the Association had a permanent national office that oversaw a national network of seventy-five chapters from every region, representing ten thousand members and the Canadian affiliate, which had established its own network of chapters.[78] Staffed by volunteers, the office answered endless requests—more than ten thousand from June 1980 to April 1981—for information and referrals.[79] The TSA newsletter reported that the October 1980 "Ann Landers" column, which gave a description of Tourette syndrome and provided the TSA address, "brought over 5,000 letters to our office in the first few weeks after it appeared."[80]

By 1980, the TSA published extensive educational materials for physicians, patients, and their families, as well as for educators and other professionals. A professional-looking quarterly newsletter had replaced the earlier mimeographed offerings. The Association persuaded prestigious advertising firms to create (gratis) professional public service announcements about Tourette's. These aired on radio and nationally televised large audience shows, including, due to the extraordinary efforts of Betti Teltscher, the "Dick Cavett Show," which, according to the April 1981 TSA newsletter, had resulted in forty phone calls a day to the national office since its first broadcast about a year before. University film and medical research departments continued to cooperate with the Association in producing films about Tourette's. Teltscher had recruited the UCLA producers of the 1979 award-winning film "The Sudden Intruder."

In 1980 the Tourette Syndrome Association had asked Arthur and Elaine Shapiro to write a pamphlet for distribution to pediatricians. By 1982 the TSA, again with financial support from the Pearl family, had printed and distributed 26,000 copies of the Shapiros' *Tics, Tourette Syndrome and Other Movement Disorders, A Pediatricians Guide.*[81] The pamphlet told pediatricians that "recent clinical studies indicate that Tourette syndrome is caused by primary organic disease of the central nervous system." But the authors admitted that this conclusion rested on circumstantial evidence: "While it is recognized that the evidence is inconclusive because a specific lesion has not been demonstrated, circumstantial evidence supports an organic etiology for tics." They also warned practitioners that because "the symptoms of Tourette Syndrome can be so bizarre, frightening and troublesome," and because afflicted children often were "disruptive in school, create interpersonal difficulties with peers, upheaval in the family, and often interfere with, or sharply limit, normal social intercourse," physicians' "usual response . . . is to recommend psychological treatment." But the Shapiros had "concluded that psychological treatment including psychotherapy is clearly ineffective as a primary treatment for tics and Tourette Syndrome." In fact, they were convinced that because "psychological studies . . . failed to support the frequently advanced notion that psychological conflicts cause tics or Tourette Syndrome" that "psychological treatment . . . for tic symptoms" was "both ineffective and inappropriate."[82]

The Association actively lobbied the U.S. Congress and other legislative bodies for regulations that would help Tourette's patients and their families gain access to drugs and receive legal protection from harassment. One of these legislative efforts resulted in extraordinary publicity about Tourette syndrome. In the late 1970s Arthur Shapiro had decided to try the experimental drug pimozide (Orap) on a young male patient who could not tolerate the side effects of haloperidol. The mother, Abbey Meyers, was pleased that pimozide controlled her son's tics, but was horrified to discover that, after Shapiro's supply was gone, no more was available because the manufacturer decided that there was not a large enough potential market to justify the twenty to fifty million dollars required for a clinical trial for Food and Drug Administration (FDA) approval. Meanwhile, Adam Seligman, an eighteen-year-old Tourette's patient in Los Angeles, who also was unable to tolerate haloperidol and for whom pimozide seemed to work, had been able to obtain the drug by bringing it in (illegally) from Canada. When U.S. Customs agents seized Seligman's pimozide at the Canadian border, he appealed to his congressman for assistance. That congressman, Henry Waxman, happened also to be chair of the House Subcommittee on Health and Environment.

At a hearing in Washington, D.C., on 26 June 1980, both Seligman and Meyers gave impassioned testimony about Tourette's and the need for federal support for "orphan drugs" like pimozide. The hearing received national press coverage, including major stories in both the *New York Times* and the *Los Angeles Times*. These accounts, particularly of young Seligman's testimony, caught the attention of Maurice Klugman, producer of the popular weekly television series "Quincy," starring Maurice's brother Jack Klugman. Maurice decided to devote an episode (4 March 1981) to Tourette's and the problems that patients like Seligman faced in obtaining potentially effective drugs like pimozide. "Quincy" not only educated the American public about Tourette's as an organic disorder, but also helped get the then stalled "Orphan Drug Bill" passed by the House of Representatives and eventually, in January 1983, signed into legislation that made it much easier for pharmaceutical firms to undertake the development of so-called orphan drugs.[83]

By 1982, the Association had begun to attract sizable donations from private individuals, corporations, and foundations to support research on Tourette syndrome and to cover administrative costs. These donations, along with the new legitimacy accorded to research on Tourette syndrome, resulted in a flood of newly funded studies, presentations at professional meetings, and published journal articles.[84] In May 1981 the TSA had sponsored the First International Gilles de la Tourette's Symposium in New York City. Almost three-hundred physicians and researchers from China, Japan, India, Britain, Belgium, France, Denmark, Germany, Canada, and the United States attended. Participants included all the major North American researchers. Medical panels discussed topics including clinical diagnosis, animal models, genetics, epidemiology, and clinical pharmacology.[85]

Figure 10.2 The Tourette Syndrome Association, a patient support group founded in 1971 by families of those with Tourette's, played a crucial role in creating public awareness of the organic causes of Tourette syndrome. Its officers, including Judy Wertheim, TSA president (left), and Betti Teltscher, former president (right), persuaded television celebrities like Jack Klugman (center) to include segments about Tourette's in their shows and later to testify at congressional hearings in support of legislation for "orphan drugs" to treat Tourette syndrome.

"The Tourette Syndrome Association," proclaimed its president, Judy Wertheim, in the spring of 1982, "is directly or indirectly responsible for all the research currently being conducted on behalf of Tourette patients. This statement may appear to be overly presumptuous, but it is not! Ask yourself who was doing research on Tourette Syndrome more than ten years ago? Who understood or cared about the plight of victims with Tourette? Practically no one!" According to Wertheim, "The endeavors of the TSA, both medical and general publicity, have made everything happen."[86] Arthur Shapiro told the tenth anniversary membership meeting in May that "we cannot discuss Tourette Syndrome without discussing the TSA, which has advanced the recognition of this disorder, educated countless numbers of educators, psychologists, social workers, and communities, and aided patients and families, contributed to disseminating information and understanding of the illness, and is now supporting research. We applaud the TSA for its part in the ongoing battle for control of this disease."[87]

Although the remarkable success of the TSA owed much to the energy of its leaders, it also drew on a wider North American historical tradition, which authorized narrowly focused grassroots movements to call professional expertise into question. In contrast, the resilience of a psychoanalytic frame for Tourette syndrome in France (no matter what its merits) reflects, at least in part, a counter historical tradition, where lay persons continued to defer to experts, especially in the arena of medicine and disease.

The Problem with Haloperidol

As Shapiro's 1982 remarks suggested, contrary to the impression that readers of the testimonials might receive, the battle for the control of Tourette's was far from over. Despite the publicity and much of the initially optimistic clinical predictions, the therapeutic results of haloperidol were increasingly ambiguous. Most of Shapiros' patients recognized that haloperidol, when it could be tolerated, acted on the symptoms of tics rather than on their underlying causes.[88] Although the Shapiros, the TSA, and the popular Tourette's narrative all cited haloperidol's suppression of tics as suggestive of a possible organic substrate for these symptoms, the Shapiros and their medical colleagues conceded that action itself was not scientific proof of an organic substrate.[89] Not until 1976 was haloperidol's action on the D2 dopamine receptor understood.[90] This discovery was impressive because it suggested that tics were connected to dopamine transmission in the substantia nigra/basal ganglia brain circuit (see figures 8.1 and 8.2). Such a finding, however, remained a correlation and still left open the question of whether involuntary tics resulted from hypertransmission of dopamine, sensitivity of dopamine receptors, basal ganglia malfunction, or some combination of all three. It certainly did not rule out the possibility that other neurotransmitters might be involved, that other brain structures might be malfunctioning, or that some other factors, such as antibody cross-reaction, might have interfered with basal ganglia function. Thus, reduction of dopamine transmission (with haloperidol) might act to control tics indirectly simply because it suppressed basal ganglia signaling. Examining some of these possibilities in 1981, Van Woert concluded that the evidence for "the dopamine hypothesis" of Tourette's was "not strong."[91] In April 1982, neurologist Richard L. Borison and his colleagues published a study that concluded haloperidol was *not* "a specific agent for treating Tourette Syndrome" because "the incidence of extrapyramidal side effects, depression, and apathy was greater in those patients receiving haloperidol" than those who were given the neuroleptics trifluoperazine and fluphenazine.[92] What the action of haloperidol had provided was a logical set of nonspecific correlations supporting the belief that Tourette syndrome was an organic disorder in those patients whose tics it suppressed; that if motor and vocal symptoms could be controlled by pharmacology, their cause must be biochemical. And, if the disorder were organic, families and patients need no longer suffer from the guilt of believing that somehow their actions brought on or exacerbated the symptoms.

Although haloperidol was clearly a major advance over other treatments—for many patients and their families it was a miracle drug—it did not work for every ticcing patient. In many cases the drug's side effects made it impossible to tolerate.[93] Clinical follow-up studies of haloperidol throughout the 1970s had presented a mixed picture of haloperidol's effectiveness. On the positive side, Richard L. Jenkins and Barry N. Fine, both at the University of Iowa, reported to the 1970 American Psychiatric Association Meeting (APA) that thirteen of fifteen ticcing and coprolalic patients whom they treated with haloperidol achieved "complete or almost complete control of symptoms" and they "had no case of Tourette's disease in which haloperidol did not reduce or control the tics." These patients had been followed for a period ranging from six months to sixteen years. At the time of the follow-up "4 patients were symptom free; 3 of these were on no medication, and 1 was on 2 mg. of haloperidol daily."[94]

Clouding the issue, however, was the fact that patients often spontaneously improved, making it difficult to be certain what role haloperidol played in an individual case. For instance, British (Essex) psychiatrist S. J. M. Fernando, writing in 1976, agreed with Jenkins and Fine that haloperidol was an effective tic suppressor, but his follow-up data presented a more confusing picture. Three of Fernando's five patients "(all women) have recovered or nearly recovered without medication. One man who improved on haloperidol continues to maintain his improvement having reduced the dosage of the drug over the years." Another "woman who improved symptomatically on haloperidol," however, "refused to continue her treatment, claiming the drug induced a 'change of personality.'" Fernando discounted this complaint, attributing the woman's refusal to continue taking the medication to her "subconscious psychological need to continue making these [cursing] utterances." But, the patient's reaction more likely meshed with Fernando's admission that "the effects on the condition of long-term treatment with the drug are not known."[95]

As early as 1972 Arthur Shapiro had concluded that although "haloperidol . . . gives almost complete relief to most Tourette's victims . . . about one-fifth of them get only around 80% relief."[96] Abuzzahab tied these failures to misdiagnoses: "If it doesn't work," Abuzzahab told a *Wall Street Journal* reporter, "the physician must question whether the patient really has Tourette's syndrome."[97] Shapiro vigorously rejected Abuzzahab's assertion, and later Abuzzahab would revise his view.[98]

As with Fernando's patient, most problematic were haloperidol's many side effects even among those patients whose tics it controlled. As the 1972 *Wall Street Journal* article reported, "There is a major drawback to the drug. Since it has to be given in large dosages, Tourette's patients starting treatment usually undergo extremely unpleasant side effects, ranging from frightening muscle contractions in the upper body to restlessness to lethargy. During the first months of treatment, while the dosage is being adjusted, to give maximum relief of symptoms with minimum side effects, patients often have to be coaxed to take the medication."[99]

Interviews with patients and their families presented a different picture from the one presented in published clinical follow-up studies. Even those patients who did not experience Parkinson-like side effects often reported that haloperidol, in the words of one of Shapiro's early patients, turned them into "zombies."[100]

University of Oregon psychiatry professor Paul H. Blachly, whose coauthored 1966 study had reported amazing success with haloperidol,[101] revised his views a decade later in a letter to Sheldon Novick. Blachly wrote that he had lost track of all but one of his original patients. The remaining case, "an elderly man who is unusually intelligent and articulate about his condition," had suffered from motor tics, barkings, unwelcome vocalizations, and "psychological humiliations" for his entire life. However, even though haloperidol had "decreased the frequency and intensity of his bark-ings, he preferred his symptoms to the dysphoric effect [anxiety] induced by the droperidol (intravenous haloperidol)."[102] Abbey Meyers, whose eldest son was under Shapiro's care, reported that although "Haldol controlled his tics, it had turned him into a zombie."[103]

Tommy, whose haloperidol testimonial appeared in *Good Housekeeping*, had begun treatment with the drug when he was nine. Over the next three years Tommy's symptoms worsened. Shapiro reacted by increasing the dosage of haloperidol. "Para-doxically," his mother reported, Tommy's "TS symptoms seemed to increase as we increased his medication. His side effects were now more alarming because he was experiencing school phobias (even though he maintained an A average), suicidal thoughts," and slow, atypical tic-like movements (possibly tardive dyskinesia) that she feared may have been brought on by the medication. Tommy's parents "had him hospitalized at the Yale Child Study Center under the supervision of Dr. Donald Cohen. Amazingly," the mother re-members, "as they detoxified him and took him off all drugs, his TS symptoms became milder and milder. Also, his suicidal thoughts vanished and he said he was looking forward to returning to school."[104]

The TSA had been receiving letters since the late 1970s from its members complaining about the deadening effects of haloperidol. One mother, an officer of the San Diego Chapter, wondered whether "squelching a child's tics completely, to make it possible for *us* to deny he had a problem, or for him to stay in public school or be on a team or in a club, is as ghastly as child-beating—it's only more subtle and more sophisticated." The decision of whether to continue her child on haloperidol was "sort of like being asked which leg we would like to have cut off. Do we want to have our child ostracized from school, sports, and friendships because of tics—or do we want him to be taken places with oatmeal between his gears instead of oil, deprived of the only tools he can really develop to create his own life out of the raw material nature has given him?"[105] As a result of these letters and her own child's ex-perience, Meyers, the Tourette's New York coordinator, alerted the TSA membership in July 1980 that "since the majority of children [diagnosed with Tourette's] are treated with Haldol, parents and teachers must be aware that this medication can add to their child's problems by blunting cognitive processes. It is mandatory that doctors be in touch with teachers and warn them of learning problems related to Haldol."[106]

In October 1980 Elaine Shapiro moderated a panel of six young adults diagnosed with Tourette syndrome. Only two were still using haloperidol. One of them, twenty-year-old Phyllis, had been diagnosed only the previous year when she was put on Haldol. Two others had switched to the norepinephrine enhancer clonidine (Catepres), a drug approved only for use to control high blood pressure. Orrin Palmer, who had been Shapiro's patient since 1965, had recently graduated from college and would soon enter medical school. He reported, contrary to the emblematic story that had appeared about him in 1975 in *Today's Health,* that he had "not been helped by any medication."[107]

By 1981, although numerous medical and popular articles hailed haloperidol as a miracle drug for the treatment of Tourette syndrome, patients, their families, and physicians increasingly expressed reservations about the drug's side effects. One of the leaders of the TSA revealed in 1981 that when in 1975 Shapiro had placed his seven-year-old child "on a low dosage of haloperidol, the symptoms improved remarkably. However several months later the symptoms returned and were more severe, and the medication dosage was increased [to] 20 mgs. of Haloperidol per day. Not only did this prove to be ineffective, but our child developed side effects, not completely recognized at the time." The parent reported that his child's problems included "difficulty in concentrating on school work, restlessness and marked problems with hand-writing. This child, who is still remembered by her first grade teacher as one of the most intellectually gifted students she had, was intellectually impaired in the second and third grades." The effects of haloperidol were so devastating that the child "had to be transferred to a private school so that there would be less pressure . . . by the teacher, who couldn't deal with her problems of restlessness and difficulty in concentrating." And, the father added, these side effects "brought ostracism socially, making for a most painful childhood."[108]

Concerns like these had become so numerous by 1983, the new TSA medical committee chair Ruth Bruun (Novick had tragically died of a brain tumor that year) wrote a column in the newsletter asking patients and their families not to overreact to reports about the potential side effects of haloperidol. "It is true," wrote Bruun, "that there has been much information about side effects of medications, particularly haloperidol, which can be frightening to parents. This information has been published in journals, the popular press and the TSA Newsletter and has been spread by word of mouth at TSA meetings." Bruun was convinced, however, that "though well intentioned, this negative information may have been overemphasized, and may actually be preventing some patients from receiving treatment which otherwise might be helpful to them." The problem, Bruun implied, may be that insufficiently experienced practitioners were unaware that "the accepted treatment methods for both Haldol and Catepres (clonidine) involve a very slow and cautious increase of medications and much attention to the development of side effects."[109]

The Shapiros never lost faith in haloperidol but had changed their minds about how to use the drug.[110] Originally, Arthur Shapiro had placed patients on high dosages and reduced them only when the first Parkinsonian tremors appeared, but by the late 1970s he advocated beginning with low dosages and increasing them until tics disappeared or undesirable side effects were evident.[111] "We have always believed," wrote Elaine Shapiro in 1998, "that haloperidol got a very bad press when in truth it has been the major effective treatment of TS. Used correctly along with judicious use of ancillary drugs, hundreds of our patients now live very fruitful lives." But she had also found that many of their patients went off medications. "They felt they didn't need it anymore, or whatever symptoms they had were controlled. And, at that point, it wasn't interfering with their psychosocial function. 'So I have a tic. You know, big deal. It just doesn't really make any difference to me. And I can control it enough so it isn't interfering. I don't think I need the medication.' And they'd go off and they did well in college and went on to productive lives."[112] Many others found that their symptoms simply abated. Others, however, were less fortunate. Neither drugs nor personal strategies proved effective. But at least, thanks to Arthur and Elaine Shapiro and the women and men who established and worked for the Tourette Syndrome Association, neither these patients nor their families would have the added burden of guilt that they were somehow responsible for their disease.

Although I have no doubt that Tourette syndrome is an organic disturbance brought about by malfunctions connected with signaling in the basal ganglia, as a historian of medicine I am equally convinced that by the 1980s the evidence for this conclusion rested more on a shared set of beliefs than on compelling scientific data. At bottom this set of beliefs relied on a series of formulaic patient testimonials that unequivocally rejected theories and interventions that seemed to blame the afflicted and their parents for the affliction. Although these narratives were an essentially accurate portrayal of numerous experiences, neither they nor the effect of haloperidol proved that Tourette's was an organic disorder. What the stories demonstrated was that dopamine antagonists and, by the 1980s, some norepinephrine enhancers could, in many cases, control symptoms, albeit sometimes with intolerable side effects. A number of theories could be constructed implicating hypertransmission of dopamine or suggesting hypersensitivity of dopamine receptors in the basal ganglia. It was equally plausible, as others had suggested, that dopamine antagonists were general pharmacological agents that, given in high enough doses, could manage any psychiatric condition.[113]

In many ways this new view of the etiology of Tourette syndrome meshed with and reflected a biological revolution that was taking place throughout psychiatry in the 1980s. In important ways, the Shapiros and the Tourette Syndrome Association were in the vanguard of that movement. Together, the Shapiros and the Association had managed to shift the burden of proof away from organic correlations onto psychogenic theory. What is astounding in retrospect is that by 1985 this shared set of beliefs—between practitioners, patients, and their families—was so powerful that the once dominant psychoanalytic psychiatric establishment, as well as behavioral psychologists, were powerless to challenge it. By then, not only was Tourette syndrome, in Meyers's characterization, "a household word," but also physicians interested in it could neither treat patients nor pursue research if they doubted or challenged the belief that Tourette's was organic. As a result it had become impossible to conduct or fund research in North America or most of Europe, except France, on the putative psychogenic factors of Tourette's. And by 1985, the Association's influence would extend to the very corridors of the Salpêtrière Hospital.

Clashing Cultural
Conceptions

II

· · · ‖ II ‖ · · ·

In May 1985 the French Neurological Society (Académie de Neurologique) hosted a two-day conference at the Salpêtrière Hospital in honor of the 100th anniversary of the publication of Gilles de la Tourette's landmark article on convulsive tics.[1] Officially cosponsored by the Société Française de Psychiatrie Biologique, Société Française de l'Enfant et de l'Adolescent, and the Tourette Syndrome Association, the idea for the meeting had been suggested almost a decade earlier by the American psychiatrist Faruk Abuzzahab[2] and urged on by the executive board of the Tourette Syndrome Association.[3] The program was arranged by Michel Dugas of l'Hôpital Herold of Paris, professor of psychiatry, who had attended the TSA-sponsored First International Tourette Syndrome Symposium in New York in 1981 and had cooperated with the Association ever since.[4] Participants included neurologists and psychiatrists from France, the United States, Britain, Belgium, and Denmark, as well as a delegation from the Tourette Syndrome Association. Among the Association leaders attending were Eleanor and Bill Pearl; Judy Wertheim, a former president; and Sue Levi, the new TSA president. Arthur Shapiro was selected as one of the three honorary copresidents of the symposium, along with the distinguished neurologists MacDonald Critchley of Britain and P. Castaigne of the Salpêtrière. The event was, in many ways, a celebration of the American resuscitation of Georges Gilles de la Tourette. Americans presenting papers—Elaine and Arthur Shapiro, Rosewell Eldridge, Thomas Chase, and Arnold J. Friedhoff—were those researchers who had been most closely tied to both the Tourette Syndrome Association and the organic view of Tourette syndrome. In contrast, none of the leading French psychiatrists who had written on Tourette's since the 1970s, including Serge Lebovici, Yves Ranty, J. Léger, Marie Olivennes, or M. Yvonneau, had been invited. Although the French participants were important figures in neurology and psychiatry, only Dugas had actually written about or researched the disorder.[5] Another presenter, Michel Gonce, formerly from Montréal and now in Liège, Belgium, had collaborated with Dugas in a paper presented at the 1981 New York meeting.[6] The British and Danish participants were advocates of an organic etiology of Tourette's. With the exception of Dugas' and Daniel Widlöcher's contribution, the Americans made most of the major clinical claims about the etiology of Tourette's. Given the absence of Lebovici and other outspoken psychoanalytic critics of the organic view, this was an elaboration of the Shapiros' contributions and the TSA's point of view. Indeed, the conference title, "Centenaire du Syndrome de Gilles de la Tourette," which even in French used the term "syndrome" rather than "maladie," revealed that this would be an affirmation of the American reading of Gilles de la Tourette's text.

The papers presented during the conference were published the following year in a special issue of *Revue Neurologique*.[7] Given the celebratory nature of the event and the participants, controversy was minimal, but important differences emerged. All of the French contributors presented evidence that, although politely embedded in their texts, questioned the assumption of the North American delegation about the effectiveness of haloperidol. They would not endorse the assertion that the etiology of convulsive tics was primarily an organic issue.

This was evident even in the opening remarks of Castaigne, director of the Salpêtrière's clinic for disorders of the nervous system, who criticized the tendency to reduce tics to their physiology. Reaching back to observations first made by Julien Rouart in 1941 and affirmed by Serge Lebovici and Yves Ranty, Castaigne insisted that "the term 'movement' applied to tic was too restrictive." Rather, he preferred the word "act" because, "as Charcot said," tics are "a significant gesture." It is not the tic's "motor phenomenon which is abnormal, but its repetition, sequence, and eventual association with other series of uninvited movements." Castaigne also questioned whether tics were actually "involuntary," as was the case in other neurological disorders such as Parkinson's, because ticcers often mimicked others and themselves and were able to restrain their tics from time to time. One could not separate the interaction of "the highest levels of sensory motor organization," and "indispensable intermediate psychic" functions. Both, Castaigne asserted, were involved in the production of tics. He hoped that looking for the interaction between these two phenomena would move the meeting beyond the old debate of psychogenic versus organic etiology. For Castaigne was convinced that the symptoms described by Gilles de la Tourette could not be explained only by reference to "hyperactivity of dopanergic systems."[8] With these brief remarks, Castaigne had set forth the central position of all the French commentators that Tourette syndrome could not be understood as a solely organic disturbance, even if physiological factors were admitted to play an important role.

Thus, in an ostensibly historical overview,[9] child psychiatrist and conference organizer Dugas[10] was not ready to eliminate a role for psychoanalytic psychiatry in either the diagnoses or treatment of the disorder. Like other French observers, Dugas saw tics, compulsive acts, and obsessive thoughts as a form of "gestural and phasic motor activity." These, he believed, were not merely organic because both tics and compulsions are often "an equivalent of what we use to communicate with the outside world and with ourselves." Like Castaigne, Dugas called for an integrated approach: "neurologists and psychiatrists, along with psychopharmacologists, physiologists, psychologists, and linguists must stand together to overcome the challenge that disease presents for the neuro-sciences."[11]

Elaine and Arthur Shapiro did not directly address the concerns of their French colleagues. However, they were unequivocal in their rejection of a central tenet of French psychiatry, first expressed by Georges Guinon in 1886, reaffirmed by Pierre Janet in the 1890s, and confirmed by almost every important French observer since, that tics were manifestations of deeper obsessions and compulsions. Based on their evaluation of 1,490 patients examined for movement disorders, the Shapiros found no "association between psychopathology and uncomplicated Tourette's disorder" and "no known predisposing precipitating or psychodynamic factors." They specifically ruled out obsessions and compulsions as features of

Tourette's. Pointing to the DSM-IIIR definition of obsessions, the Shapiros reported that "most GTS symptoms do not fulfill the criteria for obsessions." The same was true for compulsions, which according to the DSM-IIIR required "representative, purposeful and intentional behavior that is performed according to certain rules or in a stereotyped fashion."[12] These assertions were directly challenged later in the conference by Daniel Widlöcher.

Eldridge and neurologist Martha B. Denkla, coworkers at the National Institute for Neurological and Communicative Disorders and Stroke in Bethesda, Maryland, examined etiological factors. Finding no environmental or hereditary causes, Eldridge and Denkla argued that the etiology of Tourette's "most probably will be elucidated through more systematic anatomical and biochemical researches."[13] Neurologist Michel Gonce asserted that brain imaging studies revealed substantial differences between simple tics and the more complex tics produced in Tourette syndrome, adding support to the view that Tourette's symptoms resulted from "a biochemical cortical-subcortical dysfunction."[14]

Three papers examined the neurochemical mechanisms of Tourette's. Based on brain imaging studies and the action of neuroleptics, Chase and two of his colleagues from the National Institutes of Health concluded that the etiology of Tourette syndrome was most likely connected to hypersensitivity of dopanergic receptors, rather than to excessive production or transmission of dopamine.[15] The Danish team, Rasmus Fog and L. Regeur, agreed with Chase and his colleagues that Tourette symptoms more likely resulted from hypersensitivity of striatal dopamine receptors than from general dopamine overactivity.[16] Finally, Friedhoff discussed the role of the D2 (striatal) dopamine receptors in Tourette's.[17]

It was left to Daniel Widlöcher, the highly respected and diplomatically astute chief of psychiatry at the Salpêtrière, to reassert the French position without transforming the celebration into a conflagration. Widlöcher was well-tested for such a task. Trained by Jacques Lacan, Widlöcher was assigned the difficult task in 1961 of informing his mentor that he could no longer conduct training analyses at the French branch of the International Psychoanalytic Association because Lacan refused to adhere to the rule that an analytic session last one hour.[18] By the 1980s Widlöcher, often a collaborator with Lebovici, was a leader of French psychoanalysis.[19]

Widlöcher's reaction was that all organic (reductionist) claims should be placed in both historical and clinical contexts. "For the past hundred years," Widlöcher reminded his audience, "there has been a continuous debate over whether Gilles de la Tourette's disease was a mental or neurological illness." Conceding the weakness of Ferenczi's "radically psychoanalytic view" that these symptoms were a form of conversion hysteria, Widlöcher, nevertheless, was unwilling to admit that the disorder should be understood only in terms of physiology. There was no doubt that "neurophysiological mechanisms" played a role in the expression of these symptoms, "but that was also true in all mental disturbances. There is no thought produced, normal or pathological, without a specific brain state, and there is a neurophysiological mechanism of tics as there is one for obsessions or reactional depressions." The etiological issue, according to Widlöcher, was "to define the primary disturbance." "At this moment," he asserted, "we don't know what that is." Certainly, the frequency of tics during childhood required asking if the problem was as much in the timing as in the triggering of tics. "There is an indisputable familial risk. But all of this does not permit us to discard possible precipitating psychological factors or [psychological] contributions to the persistence of the symptoms. In fact, the most pertinent question is a clinical one: are the symptoms of the illness a simple expression of a neuronal connection or of an anomalous gestural expression due to this alteration?" Was the problem the movement or the abnormal gesture? To answer these questions, said Widlöcher, clinicians should not merely set up "different groups of symptoms," as if they can be "dissociated from the interpretations that we can give them."[20]

Because the mechanisms that triggered tics could be "observed in numerous neurological syndromes," Widlöcher was convinced that the specific "facts" of tics could not be reduced to or explained solely by these physiological mechanisms. Widlöcher preferred interdisciplinary approaches, particularly a recent "elegant neurological study" that argued that tics initially were spontaneous, but "those tics which persisted were imitations of the earlier tics." To demonstrate this, the authors had shown that electrical impulses of the initial tics were different from those in the later tics, suggesting that spontaneous childhood tics were different from those that developed to be part of Gilles de la Tourette's disease.[21] Thus, Widlöcher used brain imaging reports as evidence for the possible psychogenic involvement in convulsive tics.[22]

Widlöcher made a similar argument about the connection of compulsions and obsessions. Contrary to the Shapiros' assertions, he insisted that a review of most recently published American studies supported a link between obsessive-compulsive behavior and Tourette's. Using data from these articles, Widlöcher demonstrated that 60 percent of all patients identified in these studies were diagnosed as having obsessive and compulsive behaviors along with their tics. The association of tics and obsessional phenomena was too frequent, Widlöcher was convinced, to be the result of chance. The problem again was interpretation. He doubted a purely neurological explanation of the obsessive and compulsive behaviors in those with tics because obsessional symptoms were not affected by the same medications that ameliorated tics. There were many possible psychological explanations, but Widlöcher was most persuaded by those that attempted to integrate the biological with the psychogenic. He endorsed a 1983 study suggesting that neurons can simultaneously be affected by physiological and psychodynamic factors, and that compulsive ideas excited different neurotransmitters than movements, such as tics, did.[23]

Making similar arguments about other associated behaviors found in Tourette's patients, Widlöcher concluded that the disorder "could not be reduced to a motor problem similar to that of chorea," but rather it provided "a unique model which must be studied with care. We should be suspicious of easy solutions." If we used "to explain a tic by its obsession, we are tempted today to explain the obsession by its tic." Finally, Widlöcher argued that "neither neurological, nor psychiatric, Gilles de la Tourette's disease shows us again that the brain is a whole and that neurophysiology and psychopathology will only thrive through interdisciplinary approaches."[24]

Widlöcher's views were echoed broadly by British neurologist MacDonald Critchley, who warned that naming symptoms as a syndrome sometimes made it difficult to see them as possible exaggerations of typical forms of human behavior. He speculated that the onset of tics and vocalizations may parallel the way speech developed in prehistoric humans.[25]

Nevertheless, in the roundtable discussion that followed, the emphasis returned to the organic features of Gilles de la Tourette syndrome. Widlöcher's objections went unanswered and, as the local host, he was typically too polite to engage in a prolonged dispute with his American guests. Elaine Shapiro remembered the meeting as extremely pleasant.[26]

Lebovici's Response

Absent from the festivities, Serge Lebovici's reaction to the Americans was much less restrained than Widlöcher's. With four of his former students, Lebovici published a fifty-five-page response to the centenary, entitled "A Propos de la Maladie de Gilles de la Tourette."[27] This celebration, according to Lebovici, exaggerated claims of the diagnostic contributions of Gilles de la Tourette, whose work was "little known in France," although it had been widely advertised by "the Tourette Syndrome Association, created in the United States in 1971." As a result, Lebovici claimed, Gilles de la Tourette's work had assumed unmerited "importance in the Anglo-Saxon medical community."[28]

Since Freud, Lebovici recalled, childhood neuroses like these have been treated by psychotherapy, but "that no longer seems the order of the day in many countries since success with certain neuroleptics has . . . led to a hypothesis about the connection between the syndrome and brain malfunctions." This "so-called neuropsycholgical approach," according to Lebovici, had "taken the form of a syllogism" that argues that because tics and obsession can sometimes be ameliorated "by certain neuroleptic medications which act on the neurotransmitter system," they must be of "cerebral origin."[29] "This reversal [of logic] aims to support a neuropsychological version of neurotic manifestations" by "attempting to provide a psychophysiological explanation" for "psychopathologies" that in Lebovici's view were informed by "unconscious conflicts."

Lebovici believed that "the papers presented at the symposium organized in Paris . . . furnished the occasion . . . to repudiate the work of psychoanalysis, . . . returning in force to the dualistic theses which unhappily place psychoanalysis and biology in opposition." The most extreme example of this tendency was in the presentations by the U.S. participants, especially that of the Shapiros. Contrary to the physiological conclusions reached by the Americans, Lebovici believed that a careful rereading of the data in the Americans' presentations supported a psychoanalytic interpretation. Consideration of a patient's familial conflicts would reveal an underlying meaning in the apparently confused motor discharges and seemingly involuntary vocal outbursts. Because the majority of ticcers were males, Lebovici was certain that symptoms were rooted in oedipal conflicts. The irregularity of the motor movements, their suspension "when a child submits himself to a task which holds his attention," and the relief experienced when a tic is discharged provided additional evidence for the psychological impetus of these behaviors. The most persuasive evidence for a psychogenic thesis was the progression of vocalizations from explosions of noise, often to barking, and then on to scatological and obscene expressions. In many cases these began with "compulsive insults directed to the father," such as "shit on my father," followed by "sacrilegious thoughts addressed to God: 'Shit on God.'" These were followed by "attempts to cancel the insults and sacrileges by rites of beseeching, such as compulsive praying," made continually necessary by renewed insults. In addition, Lebovici noted the presence of sexual rituals, often including "a compulsion to touch one's genital organs."[30]

For these reasons Lebovici and his colleagues rejected the DSM classification of Tourette syndrome as separate from other obsessions. They were persuaded by their clinical experience of Freud's claim that Emmy von N's tics and vocalizations defended against unwanted (unconscious) ideas. This provided the starting point for understanding the underlying causes and proper treatments of maladie des tics. The authors drew on Mahler's work to exemplify Freud's view that biological and psychological issues were inseparable. From this perspective, Lebovici endorsed Widlöcher's presentation at the centenary meeting because Widlöcher had "remained persuaded that Gilles de la Tourette's disease is not purely a neurological syndrome since it is made up of gestures: the anticipatory representation of a tic shows that it is not only the expression of motor impulse." Moreover, Widlöcher believed that "this impulse retains a compulsive value in the tic" and can be treated therapeutically as if they were "obsessional symptoms." Lebovici also applauded "the conclusion of M. Dugas" that while "it is probable that neurotransmitter disturbances influence . . . the symptoms of Gilles de la Tourette's syndrome, it is not possible today to suggest a coherent biochemical explanation of this syndrome."[31]

For evidence, Lebovici cited the case of an adolescent North

African male, named Abdelaziz, afflicted with "practically incessant" multiple motor tics, barking, and coprolalia. Abdelaziz's symptoms—head jerks and shouts, including "shit, whore," accompanied with obscene gestures—first appeared when he was fourteen years old, after his leg was bitten by a dog. Abdelaziz, wrote Lebovici, was "a smart, dutiful and polite adolescent." When the patient thought of swear words, he would try to avoid saying them by jerking his head violently or sometimes by hitting himself or "by putting a finger in his mouth or by biting his shirt collar." But, "the words came out" anyway. In addition, Abdelaziz constantly kept "his hand on his genitals and pull[ed] his penis."[32]

To persuade the patient and his family to sanction the psychoanalytic therapy,[33] one of Lebovici's colleagues, Tobie Nathan, instituted a series of meetings with Abdelaziz's parents, stressing the value of "acting out" the symbolic meaning and actual events, a method that Nathan believed to be consistent with the family's cultural practices. Nathan hoped that this approach, called family "ethnopsychoanalysis," would "establish the significant lines between familial fantasies developed . . . and childhood psychic material elaborated during the psychodramatic sessions."[34]

Based on this ethnopsychoanalytic approach, Abdelaziz's examiners concluded that the boy's barking noises resulted from his "identification" with the dog that bit him. It was "at the same time an immediate defensive measure, put in place by the patient, faced with the actual aggression of the dog." The patient's "identification with the aggressor dog, which had triggered the barking tic, was followed . . . by a series of compulsive miaowing," during which he simultaneously "tormented a cat." These acts exposed "the importance of the special and powerful identification that Abdelaziz placed on the object of his anguish." The boy's identification with the dog was a symbolic substitution for "Abdelaziz's intense identification with his father." But, "despite the highly significant character of the representational material," the therapists reported that "Abdelaziz had only weak access to its symbolic content."[35]

In particular, Abdelaziz resisted the suggestion that he had any sexual desires. The fifteen-year-old continued to employ a child's vocabulary in sexual matters. For instance, he used the childhood word "zizi" referring to a penis. Although the therapists recognized that in Islamic families sexual discussion was often repressed and restrained, they nevertheless were convinced that Abdelaziz's resistance to discuss sexual issues was more extreme than that normally found among those with his cultural background. The patient's explosive foul language and his symbolic touching of his sexual organs, when combined with his reticence to discuss sexual issues, were evidence for Lebovici and his colleagues of the relationship between the sexual repression imposed by the father (claiming adherence to the rules of God) and the particular manifestation of Abdelaziz's symptoms: "In fact, Abdelaziz was afraid at the same time of the dog, of his father, of the threat of external castration, as he struggled, by his defensive counter-investments and his reaction formations, against the intensity of his sadistic fantasies." Faced with his repressed memory of witnessing his parents during sexual intercourse, "the primal and sadistic sexual scene," Abdelaziz "was only able to scream compulsively 'bitch' [salope] or 'zob, zob, zob,' identifying himself with the devouring dog or the father who hits him."[36]

After eighteen months of psychodrama Abdelaziz's shouting attenuated, but, his therapists admitted, it reappeared under stress. Lebovici's team concluded that Abdelaziz's case demonstrated the role of emotional tension in exacerbating tics in patients with powerful unresolved childhood sexual conflicts. Moreover, the case proved that coprolalia always has a symbolic signification, whose meaning must be exposed if the underlying causes of the stress are to be uncovered. "The child ticcer lives in a family which is . . . often irritated by the tics" even though the family is a contributing factor in the production of these tics.[37]

Lebovici had included an extensive discussion of this case to show that tics could only be cured if the family was included in the therapeutic process, which, as in the case of Abdelaziz, might require therapists to adjust their interventions to the cultural values and practices of that family. The problem with pharmacological interventions was that they ignored both meaning and cultural variations. Biological psychiatry failed to uncover the psychodynamic underpinning of these symptoms.

Returning to the central theme of his article, Lebovici wrote that "this neuropsychological conception . . . seduces those who want simple solutions to complex problems," especially the families of those afflicted. He pointed to the role of the American Tourette Syndrome Association, which exclusively supported psychopharmacological research in an attempt to eliminate "all feelings of culpability among the parents." Although Lebovici hoped that effective drugs would be found to cure tics, he insisted that patients' life histories were "inscribed in their disorder." Psychoanalytic insights not only would "considerably ameliorate the effects of the chemotherapy," but also could help patients explore the meanings of their tics and thus, their lives.[38]

Events in the United States

While Lebovici's view continued to inform the thinking of French psychiatry into the 1990s, the Shapiros' influence among North American researchers declined. Ironically, the inclusion of obsessions and compulsions in Tourette syndrome, an issue that French psychiatrists had raised as central to their critique of the Shapiros' organic paradigm, would play an important role in the deterioration of the Shapiros' influence, but in a context that would hardly satisfy the criticism of Widlöcher or Lebovici.

By the time of his death in 1995, Arthur Shapiro had concluded bitterly that he had become isolated from the Tourette's research community and the Association he had been instrumental in establishing. Elaine Shapiro remembered that by the mid-1980s, after returning from their leading role in the centenary celebration, they felt increasingly unappreciated by the Tourette Syndrome Association. "We . . . felt betrayed . . . There were more and more international conferences. Conferences which drew lots of people. He was always given a very minor role [by the TSA]. It was a terrible, terrible time for him. He just felt that he had done so much and had moved the field forward so much and not been given recognition that he felt he had deserved."[39]

Exacerbating these feelings was that author and neurologist Oliver Sacks rather than Shapiro was credited by the media for having brought Tourette syndrome to public consciousness. Shapiro contrasted his own careful clinical work with Sacks's idiosyncratic and anecdotal approach to a clinical investigation.[40] Sacks's description of a Tourette patient in his 1987 book, *The Man Who Mistook His Wife for a Hat,* received more publicity than anything Shapiro had ever written.[41] Sacks's 1992 *New Yorker* magazine article, detailing the tribulations of a Canadian surgeon with TS, cemented his position as Tourette syndrome's most prominent medical expert and celebrity.[42] "Oliver Sacks," Shapiro remarked in 1994 with some irritation, "is a much better writer than he is a clinician."[43]

Shapiro's personality was another reason for his eventual isolation. Arthur Shapiro could be stubborn and dogmatic; he also could be brilliant, committed, and an exciting partner and colleague.[44] However, his stubbornness, compared with the more conciliatory views of earlier haloperidol researchers, had proved essential in identifying psychoanalytic psychiatry as the "enemy" and in establishing the diagnosis and treatment of Tourette syndrome as an organic rather than a psychogenic illness. Ironically, Shapiro's increasing isolation could be attributed to the factors that earlier had accounted for his enormous influence. The decline of his influence cannot be separated from the increasing frustration that other physicians, patients, and their families felt about the effects of haloperidol and other tranquilizing agents. By 1982 the Association lay leaders and medical researchers unanimously felt that TS research should focus more on etiological issues. Although Shapiro did not disagree, a number of his strongly held assumptions about Tourette's isolated his participation in this research.

First was the issue of phenotype—what should or should not be seen as a symptom of the disorder. From the beginning, Shapiro had been adamant in refusing to include obsessive and compulsive symptoms as part of Tourette's. In the 1960s and early 1970s this refusal made political sense because obsessive-compulsive behavior was thought to be a functional (psychological) problem. But by the mid-1980s, the majority of families and medical researchers were convinced that these behaviors were an integral part of Tourette's. As drugs that increased serotonin transmission began to prove successful in controlling these symptoms, obsessions and compulsions also were moved out of the psychogenic and into the organic category. Shapiro, however, remained unyielding on this issue and, if anything, his opposition to inclusion of obsessive symptoms grew stronger.[45] He was certain that any data collected in support of obsessive behaviors in the patient population was the result of an ascertainment bias: Those with the most florid and greatest combinations of symptoms (comorbidity) tended to end up in psychiatrists' offices, especially if their offices were in research hospitals.[46] Shapiro's final published work on Tourette's (1992) focused on this issue.[47] By then almost every major Tourette's researcher, including his former resident and collaborator Ruth Bruun, had rejected Shapiro's point of view.[48] Even Elaine Shapiro, who publicly remained loyal, had disagreed privately with her husband about this issue for more than fifteen years.[49]

Shapiro's position on compulsions had clinical implications for patients with these symptoms. One result was that increasing numbers of Tourette's patients who believed that obsessions and compulsions were part of their illness sought treatment elsewhere. As important, the Association included obsessive behaviors as part of Tourette's and referred patients to those who agreed. This conflict also made it difficult for Shapiro to compete for research funding because his definition of the disorder no longer meshed entirely with that of the now powerful Tourette's community.

Second, the increasing focus on etiological issues turned toward genetics after the mid-1980s. Although the Shapiros did not discount the possibility that genetics played an important role in Tourette syndrome, they were skeptical that it alone held the key to the etiology of TS. In the early years, Association leaders appeared to accept Arthur Shapiro's skepticism, though Sheldon Novick, the scientific adviser, always believed genetic factors might be discovered. Subsequent medical research, especially uncovering the genetic marker for diseases like Huntington's, had converted many in the Association at the very time when investigations reported increasing evidence of familial patterns of Tourette's. At first, Shapiro's resistance had wide support because the initial publications on genetics by David Comings at City of Hope relied on a much wider definition of the phenotype than most in the Association or medical community were willing to tolerate. For instance, Comings included alcoholism and conduct disorder. Yet the idea of an errant gene as the cause of putative hypertransmission or, more likely, hypersensitivity of dopamine receptors made sense. Although they disagreed with Comings's phenotype, David Pauls and James Leckman at Yale confirmed a number of Comings's insights.[50] Increasing numbers of families reported discovering multiple Tourette's victims among their close relatives, including many who, like the Pearls, were influential members of the Association.

Shapiro believed that many of these genetic claims rested on questionable data and flawed logic. Like the issue of obsessive-compulsive behaviors, he noted that the data on familial patterns also resulted from an ascertainment bias, made worse by inclusion of obsessive-compulsive patients as part of the TS phenotype. By the late 1980s and into the 1990s, the Association encouraged research that focused on possible genetic issues, adding to Arthur Shapiro's exclusion. Elaine Shapiro believed that the Association's focus on genetic research was an important reason for their increasing separation from the Association. "We have always thought that the amount of emphasis on the genetic was too much," she explained. "We felt that the emphasis and funding of genetic research may have resulted in a disemphasis or neglect of other areas of research that were viable." By the end of the decade, the Association dropped the Shapiros from the scientific research board.[51]

In the 1970s Arthur Shapiro had considered the possibility that Tourette syndrome was connected to an infectious substrate. He was conversant with the long history and evidence connecting ticcing behaviors with a number of sequels to infection, including encephalitis. Certainly, given his extensive knowledge of this history, he was extremely well-prepared to follow up this line of research. He had encouraged a young medical student, Frank Arena, in his proposals for such research at TSA meetings in the early 1970s. But for a variety of reasons, including his positions on the phenotype and genetics, Arthur Shapiro never followed this possible link with Tourette's. Others who in the 1990s would systematically explore the link between antibody reaction to Group A beta hemolytic streptococcus and Tourette's, therefore, were unaware of and uninfluenced by Shapiro's earlier insights. And, unlike Shapiro, they would include obsessive-compulsive disorders as part of their phenotype.

By the end of his life, Arthur Shapiro, who, with Elaine Shapiro, initiated a revolution on the conception and treatment of ticcing behaviors, had, like many other pioneers, become separated from the fruits of his victory. But also, like so many pioneers, the reason can be attributed to the factors that made him a revolutionary in the first place. In this case, the belief that Tourette's was an organic disorder required an interpretation of the effect of haloperidol as evidence of the circular claims of psychoanalysis. Though a necessary step, it proved insufficient. Ultimately, the limitations of haloperidol exposed the necessity for other research and therapeutics. For that stage a revolutionary personality like Arthur Shapiro's was no longer required.

Research in France

French psychiatrists, too, continued to pursue the link between compulsive disorders and involuntary tics and vocalizations. They

did so, however, in the context laid out by Lebovici and his colleagues that obsessions and compulsions were psychogenic, and
that tics, no matter what their enabling organic substrate, were
also manifestations of repressed psychological conflicts. In a 1990
review article aimed at a wide range of health professionals, psychiatrist J-P. Guéguen reported that those diagnosed with Tourette's
always displayed obsessional neuroses. Despite current explanations of dopamine involvement and genetics, Lebovici's descriptions
of 1951 and 1986 held: Tourette's remained a psychopathological
disorder, an elaborate defense mechanism that reacts to some early
event in the life of the child and is quickly transformed into obsessional behavior. No matter what the treatment, Guéguen
claimed, the symptoms never completely disappear. Therefore, "we
must conclude that a tic remains a psychomotor mechanism, whose
dual function is to discharge and retain, never ceasing to discharge
the obsessional problem."[52]

By the mid-1990s, criticism of what was characterized as the
American construction of Tourette syndrome intensified and began
to take on nationalistic and almost conspiratorial overtones. Attempts to explain the etiology of tics in organic and nonpsychoanalytic terms would be met with charges of anti-Semitism.[53] In
1993, for example, Lebovici's colleague, psychoanalyst Lucien Israël, made an explicit connection between Tourette syndrome and
the Nazis in his preface to the republication of Henry Meige's 1893
doctoral thesis describing the so-called Wandering Jew Disease.[54]
Meige's explicitly anti-Semitic analysis, including numerous racist
photographs, was originally published in *La Nouvelle Iconographie de la Salpêtrière* of which he was managing editor, and
whose editor-in-chief was Charcot.[55] Gilles de la Tourette also
served on the editorial board. Meige and Feindel's book, *Les Tics
et leur Traitement,* which explained the genesis of tics in terms of
hereditary degeneration, became the most influential twentieth-
century statement about the behavior. Israël retraced these connections and, by implication, attached them all to the emergence of
the diagnosis of Tourette syndrome in America in the 1970s:

> Let us recall the curious evolution of "maladie" des tics. Was not
> Gilles de la Tourette one of the editors of *La Nouvelle Icono-
> graphie de la Salpêtrière,* of which Meige was the managing edi-
> tor? This maladie de Gilles de la Tourette, that hardly had any
> historical interest until recently, has known a veritable resurrec-

tion in the United States, under the impetus of [Oliver] Sacks (author of *The Man that Mistook His Wife for a Hat*). A fine clinician and a man of a great generosity, [Sacks] attracted attention to this syndrome by discovering three [sufferers] in only one day, just by strolling the streets of New York. The "American Tourette Association" is founded on the same basis as that of Alcoholics Anonymous, obesity anonymous, or emotionals anonymous. This is the perspective into which we should place "Meige's Wandering Jews" or, more precisely, Meige and Charcot's. Making human misery into an illness reveals a great deal. Hitler and Mussolini, not to mention other lesser dictators, were paranoid. Everyone else is healthy . . . It is the racists who are sick.[56]

In contrast to North America, the Association Française de Troubles Obsessionelles Compulsifs et du Syndrome Gilles de la Tourette (A.F.T.O.C.), established in the late 1980s, was run until its demise in 1997 by Marc Lalvée, an elementary school teacher in the northern industrial city of Lille, from his parents' house. Currently, with help from the TSA and the Norwegian Tourette Association, an American woman living near Paris is attempting to resurrect a French support group.[57] Of the five directors of psychiatry of the major medical schools of Paris, none saw any positive value in A.F.T.O.C.; most characterized the association as both ridiculous and dangerous. A.F.T.O.C.'s only serious medical support came from an allergist, Dr. Pierrick Hordé, who writes popularized tales of bizarre medical conditions and has no connection with any French medical school or research institute.[58] In the most favorable of circumstances, voluntary lay associations like A.F.T.O.C. face an uphill battle in France. Without professional connections to acknowledged experts in the field, such an association seems doomed to remain on the margins of French society.

As if to underline this position, one of France's most famous psychiatrists, Cyrille Koupernik, recently wrote that "I was astonished to learn that there were 30,000 members of the Tourette Syndrome Association in the United States; I see this as analogous with the multiple personality phenomenon [in the United States] and I am not certain that it is the most hopeful evolution of a psychiatry which has so much influence in the rest of the world as that of your great and admirable nation."[59]

Views such as those expressed by Israël and Koupernik and the

hostility of French psychiatry to the establishment of patient support groups meant that the good feelings established at the centenary celebration could not be translated into any concrete Franco–
Anglo-American dialogue on Tourette syndrome. Although by the
1980s most North American psychiatrists saw obsessive and compulsive symptoms as part of the Tourette's phenotype, they rejected
the French view that these behaviors were psychogenic. Literally
speaking different languages, the French and North Americans simply ignored one another.

The importance of the connection of linguistic issues to the debate over whether obsessions and compulsions were organic or psychogenic was evident in a doctoral (psychiatric) thesis written by
Sophie Caloone at the Université de Rennes in 1995.[60] Based on a
detailed analysis of four patients and their families, Caloone had
set out to discover "the nature of the connection between obsessive-compulsive symptoms and Tourette's."[61]

There were, she stressed, three central hypotheses in "the Anglo-Saxon consensus": First, that there is "a common genetic basis
of GDT [Gilles de la Tourette's syndrome] and certain forms of
TOC [trouble obsessionelle compulsifs]." Second, that "the severity
of the syndrome was influenced by biological factors." Finally, that
"the symptoms exacerbated or ameliorated from time to time as a
function of the quality of the internal and external worlds of the
patient."[62]

While Caloone found this "hypothesis . . . very seductive," she
agreed with the "concerns expressed by its critics" that "it rested
in the first place on a question of translation" because "the An-
glo-Saxons have translated Georges Gilles de la Tourette's word
'préoccupation' as 'obsession.'" Thus, Caloone felt, they have confused Gilles de la Tourette's discussion of the Marquise de Dampierre's *preoccupation* with her "inability to master her coprolalia
as if she were *obsessed* with it: 'The more she seems revolted by
[the obscenities'] grossness, the more she is tormented by the fear
that she will utter them, and this preoccupation is precisely what
puts them at the tip of her tongue where she can no longer control
it' [plus les mots prononcés étaient vulgaires, plus elles semblait
tourmentée par la peur de les proférer]." According to Caloone,
the marquise's concern about her coprolalia did not fit the definition of obsessive-compulsive disorder because the "DSM IIIR stipu-
lates that TOC must not result from a subordinate feature of an-

other symptom" of Tourette's.[63] Using this criterion, Caloone found that although only one of her four patients strictly fit the American definition of TOC, "an obsessional mood dominated in the family of each patient." More important, she found that patients with motor and vocal tics rarely came for medical treatment unless they also experienced other severe symptoms, such as obsessions and compulsions.[64]

For Calonne, Tourette's was not a neuropsychiatric condition in the sense that Huntington's or Parkinson's was. She conceded that Tourette's abnormal movements had a neurological substrate, but also insisted that predisposed patients fared better or worse depending on psychiatric features, not least of all familial tensions. Moreover, "certain psychiatric illnesses such as anxiety are simultaneous with GDT." Caloone, like other French observers building on Lebovici, concluded that the separation of organic and psychogenic causes of Tourette's was artificial and misleading.

The one exception to this French view was M. Dobler-Thierry's 1995 "Memoire pour le Doctcur en Science de Psychiatrie," written under the direction of Michel Dugas. Like Dugas, the French organizer of the centenary celebration, Dobler-Thierry was conversant with and sympathetic to the Anglo-North American view that Tourette's was an organic disorder. His thesis reviewed 150 patients, 54 of whom were diagnosed with GTS, 60 with obsessive compulsive disorder (OCD), 11 with a combination of OCD and motor tics, and 25 with motor tics and vocalizations. Dobler-Thierry showed how each condition implicated a particular neurobiological or neurochemical substrate. In the case of Tourette's alone, Dobler-Thierry was persuaded that malfunctions in the basal ganglia were most likely involved, perhaps as a result of genetic malfunctions. Nevertheless, like his mentor Dugas, Dobler-Thierry concluded that "many unknowns remained." He wondered whether "the twenty-first century will bring with it an answer to the relation between movement and thought, body and mind, and also a solution to their disturbances?"[65]

Whether Caloone or Dobler-Thierry represents the future of French psychiatric thinking about Tourette's remains as uncertain today as when, more than a century ago, Georges Guinon challenged Gilles de la Tourette's typology for not including obsessional and compulsive behaviors. In contrast, in North America the issue was no longer *whether* Tourette syndrome was an organic disorder, but rather to what extent the organic disorder could be traced to genetic, neurochemical, or even immune malfunctions. Although some lip service continues to be given to the usefulness of psychotherapy for amelioration of the effects of living with Tourette syndrome, there no longer is any serious research or treatment in North America that claims that the symptoms are psychogenic. Nevertheless, much of clinical value can be learned from this century-old debate about whether the etiology of Tourette syndrome is organic or psychogenic.

Clinical Lessons

12

The history of Gilles de la Tourette syndrome has important clinical implications or lessons for current practice and future medical research. First, *the disorder described in 1885 by Georges Gilles de la Tourette and subsequently named "maladie des tics de Gilles de la Tourette" by Jean-Martin Charcot is not the same thing as what today is called Tourette syndrome.* In the 1880s Charcot and Gilles de la Tourette separated "variable choreas" or "convulsive tics" from other movement disorders, including Sydenham's chorea, transient and chronic tics ("habit tics"), and obsessive-compulsive behaviors. What Charcot called Gilles de la Tourette's tic *disease* was not a *syndrome,* but an illness with a specific progressive course and hereditary (degenerative) pathology.[1]

Gilles de la Tourette's typology was almost immediately challenged by two of his Salpêtrière contemporaries, Georges Guinon and Édouard Brissaud, who rejected the construction of Gilles de la Tourette's disease as separate from other choreas and hysteria. They argued that the same symptoms—including coprolalia and echolalia—were found in both hysterics and choreics.[2] As a result of these and other criticisms, Charcot and, to a lesser extent, Gilles de la Tourette refined their claims. They conceded that motor and vocal tics and coprolalia were also symptoms of hysteria, and, therefore, these symptoms and signs were not unique to Gilles de la Tourette's disease. Charcot and his student Jacques Catrou distinguished Gilles de la Tourette's disease from a hysterical illness with similar symptoms by whether or not the *cause* was hereditary and whether or not the symptoms could be eradicated. If the patient had a degenerative family history and no intervention (such as hypnosis) was able to cure or, at least, ameliorate the course of the disease, a diagnosis of maladie des tics was justified. If, on the other hand, the patient improved and his or her family history was inconsistent with that laid out by Charcot and Gilles de la Tourette as typical of maladie des tics, a differential diagnosis of hysteria was appropriate.[3] For this reason, Freud, in the late 1880s, was confident that his diagnosis of hysteria in the case of the ticcing Frau Emmy von N was consistent with Charcot's teachings.[4]

By 1907, Henri Meige and E. Feindel's immensely influential *Tics and their Treatment* argued that only a minority of ticcers fit Gilles de la Tourette's description.[5] Tics and involuntary vocalizations, according to Meige and Feindel, resulted from uncorrected infantile habits in a population with hereditary weakness. Their view, and not Gilles de la Tourette's, supplanted all others, remaining extremely influential until the 1960s. Meige and Feindel's interpretations and cases also provided the raw material for Sandor Ferenczi's psychoanalytic view, which found its fullest expression in the 1940s work of Ferenczi's student, the American analyst Margaret Mahler.[6]

Throughout the nineteenth and early twentieth centuries, another explanation for tics—that they were a sequel to rheumatic disease—persisted in medical discourse, particularly outside the Salpêtrière circle. This view gained force in the 1920s as evidence mounted that tics resulted from postencephalitic sequels. Thus, German observers like Erwin Straus argued that the symptoms described by Gilles de la Tourette were organic rather than psychological.[7] Throughout the 1920s and 1930s psychoanalysts met this challenge to their assumptions about the etiology of tics by classifying those symptoms resulting from infection as different from those resulting from repressed childhood sexual conflict.

Because the organic substrate of tics seemed so evident, it became increasingly difficult for psychoanalysts to argue that tics were purely psychogenic. Thus, in the 1940s Mahler and her associates conceded that there was an organic predisposition for convulsive tics, but they insisted that tic symptoms only manifested themselves in those with severe unresolved psychosexual conflicts. In actual practice, however, this psychoanalytic interpretation of a psychosomatic disorder authorized treatment that focused on putative unconscious conflicts.

This historical review reminds us that almost no one in the period 1885 to 1968 accepted Gilles de la Tourette's typology. Arthur and Elaine Shapiro attacked Mahler's psychoanalytic model and in its place resurrected Gilles de la Tourette's classification system, based on the effectiveness of haloperidol on presenting symptoms.[8] But, in doing so, the Shapiros isolated Gilles de la Tourette syndrome from those symptoms associated with other movement disorders, including obsessive and compulsive behaviors and Sydenham's chorea. The latter in 1956 and again in 1976 had been identified as a postinfectious antibody cross-reaction in the basal ganglia.

From a clinical point of view there is no compelling reason to adhere strictly to the classification system erected in 1885 by Charcot and Gilles de la Tourette. There are a variety of possible underlying conditions that may result in the symptoms described by Gilles de la Tourette. Syndromes are merely convenient classification systems, and rigid adherence to them may restrict the ability to identify the underlying factors that lead to a particular condition or behavior.

The artificial separation of Tourette syndrome, Sydenham's chorea, and compulsive behaviors into distinct disorders was contested throughout much of the nineteenth and twentieth centuries and was not always supported by clinical evidence. Although medical commentators throughout the nineteenth century observed clusters of symptomatic differences that might justify separating each form of chorea as a distinct disorder, they were reluctant to do so because of a widespread belief that most, if not all, choreas were connected with a prior attack of rheumatic fever.[9] In the nineteenth century, the term "chorea" was generally used by physicians to describe a variety of highly variable movement disorders that included chorea, tics, and muscle spasms, as well as those behaviors called "variable choreas" or "convulsive tics." Variations in movement symptoms were seen as different manifestations of a common underlying condition. In contrast, Charcot rejected a rheumatic connection to Sydenham's chorea.[10] He also insisted that Sydenham's chorea was distinct from symptoms identified by Gilles de la Tourette in 1885. Yet, in terms of underlying hereditary pathology, Gilles de la Tourette's disease and Sydenham's chorea were, according to Charcot's description, indistinguishable.

Unlike Charcot, most European and North American physicians continued to assume a link existed between prior rheumatic disease and subsequent tic behaviors.[11] Moreover, building on the new science of bacteriology, a number of late nineteenth-century physicians believed that a common infectious mechanism underlay all these movement disorders, and thus refused to separate chorea and convulsive tics into distinct disease categories.[12]

In 1956 Angelo Taranta and Gene Stollerman established the connection between the onset of Sydenham's chorea and prior infection by Group A beta hemolytic streptococcus (GABHS).[13] Two decades later Gunnar Husby and his colleagues implicated the basic mechanisms of antigen antibody response (molecular mimicry), in which antibodies released to attack invading GABHS also attacked similar sites (called epitopes) in the body's own cells—in the case of Sydenham's chorea, cells in the basal ganglia area of the brain. This antibody reaction thus interfered in some way with the control of motor movements regulated in the basal ganglia.[14]

A similar pathological substrate with that identified in Sydenham's chorea would seem to make sense in explaining the manifestations of Tourette syndrome, transient and chronic tics, and obsessive behaviors in *some* patients. Recent research has implicated GABHS as possibly providing the environmental trigger in genetically susceptible families for a variety of movement disorders including Tourette syndrome.[15] This is consistent with the implication of hypersensitivity of dopamine receptors in the basal ganglia and its associated pathways as the possible neurological substrate in Tourette syndrome behavior.

If alteration of the basal ganglia by immunologic cross-reactive antibodies forms a common factor for some patients who manifest one of a number of movement disorders, including Sydenham's, Tourette's, and obsessive-compulsive behaviors, the assumption that these are separate diseases defined by a peculiar set of symptoms must be questioned. Moreover, such a view suggests a variety of alternative therapeutic interventions, a number of which already have been attempted at the National Institutes of Mental Health, including the use of prophylactic antibiotics, plasmapheresis, intravenous immunoglobulin (IVIG), and prednisone.[16] Exploring these hypotheses should not be constrained by classificatory boundaries erected in the late nineteenth century, boundaries that even at that time were contested by numerous researchers and much clinical evidence.

Although they should not be misled by artificial classificatory boundaries, *researchers should remain extremely skeptical about*

new claims and interventions, even when they appear to "work."
Even though I remain personally involved with one research team
investigating the possible role of molecular mimicry, the historical
record examined in this book suggests both caution and skepticism
toward any claims that the etiology of tics can be reduced to a
single cause. In 1825 Jean Itard was certain that his patient Mme
de C. developed tics and barking because she had refused to accept
her role as a wife and potential mother. Itard was convinced that
he had cured Mme de C. through public shaming and insisted that,
had he been given a similar opportunity, he would have rid the
Marquise de Dampierre of her tics and cursing by a similar intervention.[17] Of course, Itard never reported whether Mme de C.'s
"improvement" was sustained.

Gilles de la Tourette and Charcot solved the problem of justifying
their theories by claiming that those patients who recovered were
hysterical, and thus did not actually suffer from maladie des tics,
but those with "real" convulsive tic illness were degenerates with
no hope of a cure. Again, follow-ups were rare and, in any case,
of only a very short duration. Brissaud and his students, Meige
and Feindel, were convinced that tics were habits that took root
in those with infantile personality, who having experienced an earlier actual irritation continued to react to it even though the actual
cause had long since disappeared. Their solution was "reeducation"—a combination of exercises, physical constraints, and aver-
sions. They too claimed a high therapeutic success rate; failures
were ascribed to the unwillingness of patients to follow the reduc-
tive course.

Ferenczi and his followers were certain that tics and coprolalia
resulted from repressed masturbation and were equally convinced
that psychoanalysis and educating parents achieved great therapeutic success. Others, like Straus in 1927, tied these behaviors to a
postencephalitic cause. Their solution was treatment of the infection.[18] When tics persisted, Straus and his followers assumed the
infection had done irreversible damage. On the other hand, there
was Laurence Selling, whose adherence to focal infection theory
led him to remove the sinuses and tonsils of ticcing children, again
with claims of substantial cures.[19]

Mahler and her followers conceded that an organic substrate
was necessary, but not sufficient without psychic conflict to set off
tics. Their solution was to treat the tic symptom psychoanalytically,

focusing on the childhood conflicts that putatively activated the organic potential. When they failed to rid children of tics, Mahler and her colleagues adopted explanations that reinforced rather than undercut their theoretical stance.

In the mid-1960s Langlois and Force argued that tics and coprolalia were organic in genesis and, given the role of infection, on a continuum with Sydenham's. Treating their patient with neuroleptics, anti-inflammatories, and antibiotics, they too, with no published follow-up, claimed to have unlocked the door to these behaviors. The Shapiros and their colleagues did publish extensive follow-up studies that reinforced their claims that haloperidol was an effective drug for the suppression of tics. But, as many of their initial patients later discovered, the side effects associated with haloperidol and pimozide were often more difficult to tolerate than the symptoms themselves. Moreover, Arthur Shapiro had insisted on a definition of Tourette syndrome that excluded comorbid symptoms like obsessive-compulsive behaviors. As a result, the outcome of treatment was measured by the effectiveness of an intervention in suppression of tics and vocalizations, but not other possibly related behaviors.

Almost all claims of "successful' interventions were reported after a very brief follow-up period, which, given the waxing and waning as well as unpredictable nature of the disorder, is inappropriate. Anyone who has followed a group of Tourette's patients over time realizes how tentative all interventions tend to be. Given the history of "totalistic claims" about the causes and the "cures" of Tourette syndrome, many of which reflected the best medicine of their time, a skeptical perspective must be maintained about even the most scientifically informed current claims. This is crucial, if only because the history of Tourette syndrome is filled with so many persuasive theories and treatments, most of which have made it difficult for insights and research that contradicted that day's paradigmatic certainties to emerge.

These changing claims of the etiology of Tourette syndrome often were due less to compelling and robust scientific findings than to the dynamics of the political culture of medicine. The difference between French and Anglo–North American medical views about Tourette syndrome reminds us that medical research and clinical findings, in themselves, often are insufficiently persuasive in the face of long-held assumptions about the etiology and treatment of syndromes. Rather, as with the case of Tourette's, a successful challenge to a long-standing medical hypothesis requires not only new and robust clinical findings, but also the cultural, political, and economic support from the afflicted and their families.

The rise and fall of each successive explanation for and treatment of Tourette syndrome has been as much a story of the power of a shared set of beliefs of a professional faction as it has been a vindication of either rigorous scientific testing or carefully analyzed clinical results. Often, as in the case of Margaret Mahler's psychoanalysis, the failure of an intervention to achieve amelioration of symptoms was projected onto the afflicted and their families, while the assumptions of the theory itself remained unquestioned. This was no less true of a number of interventions that assumed that involuntary tics and vocalizations were organic in origin. Therefore, advocates of focal infection theory insisted, despite meager follow-up data, that removal of sinuses and tonsils cured tics.[20] Still others who performed irreversible surgical procedures on patients' brains made similar claims, ignoring the great collateral damage they had inflicted and the impact on the patient's overall cognitive abilities.[21]

Many of these interventions were sustained because patients and their physicians, both desperate for results, mistook the natural waxing and waning of symptoms for therapeutic success. The power of the placebo effect, whereby a medical procedure in and of itself seems to authorize a patient's improvement, should never be ignored. Unfortunately, the history of Tourette syndrome suggests that placebo improvements have been of extremely short duration for most patients, while the collateral effects of a number of questionable interventions have been permanent.

The clinical lesson to be learned from all of this is that the rise and persistence of a particular theory for and intervention in Tourette syndrome often reflected wider enthusiasms in medical practice rather than robust scientific evidence. This result was not always internal to medicine. One example of how powerful external forces have framed the understanding of Tourette's symptoms is found in the case of postwar French resistance toward organic explanations. Another is the power of patient support groups, reflecting a long tradition of voluntary associations in America. The triumph of the organic view of Tourette syndrome most likely would have been less total and longer in coming were it not for the efforts of the Tourette Syndrome Association.

Clinical claims, it seems, cannot be separated from either culture or politics. As many researchers continue to learn, support for research into new and promising arenas requires more than scientific and medical skills. Given this reality, the best allies may be patients and their families. The afflicted have as much to tell us about their disease as researchers and clinicians have to tell one another. *Practitioners and researchers can gain much from listening to what their patients and families tell them.*

If her doctors had listened to the Marquise de Dampierre, rather than appropriating her symptoms to shore up their preconceived theories, much suffering may have been ameliorated. Instead, many subsequent physicians concentrated on the patient's "will," ignoring the repeated testimony by the afflicted that they were unable to control their symptoms (at least for more than a brief period).

This was not merely a problem found among nineteenth-century observers of tic. Meige and Feindel's influential twentieth-century text appropriated the "confessions" of O. to demonstrate that ticcers could not be trusted. Throughout the twentieth century, those who treated ticcers (by almost every method available) rarely heeded their patients' complaints that their problem was their tics. Ironically, it was the psychoanalysts who appeared to pay the least attention to their patients' words. Certainly this was exposed by Mahler's eleven-year-old patient Pete, who, frustrated that his therapist refused to see his tics as his central problem, proclaimed that psychoanalysis was "witchcraft."

Even today, influenced by the success of pharmacological agents for symptom control, practitioners often neglect to listen carefully to the particular experiences that patients and their families report. "I often wish doctors and researchers would leave the comfort of their laboratories and offices to spend time at a local TSA support group meeting," writes Susan Hughes, a mother of an afflicted child. "I believe they would gain more knowledge from listening to parents and hearing the family histories than they would learn from drawing another tube of blood, ordering another EEG or completing another questionnaire."[22]

Certainly, researchers are closer to understanding the organic basis of the symptoms labeled Tourette syndrome. But even as more is learned about how genetic and postinfectious factors may combine to produce these symptoms, patients and their families will still require the empathetic skills of their practitioners. If history has taught any lesson, it is that the afflicted also have something to tell us about their affliction. What they often try to tell us is that a person is not a disease. "The medical and mental health communities are venturing to make sense of Tourette syndrome," wrote Adam Ward Seligman and John H. Hilkevich in their 1992 collection of essays. "For those of us afflicted with it and those who live and work with us, making sense of it may be less important than finding ways to use it, even to be empowered by it. That is what kept . . . thousands of others alive in the Nazi concentration camps where it was much easier to let go and die. We can integrate and even be grateful for suffering and pain that makes our lives better than it would have been without the trials and tribulations."[23]

Both Seligman and Hilkevich suffer from florid cases of Tourette's; their brains curse despite their attempts to suppress the words. Tourette syndrome is most debilitating when its symptoms persist and when persistent symptoms include echolalia, blurting out inappropriate phrases, compulsive touching, and, most of all, involuntary cursing. Of course, tics and vocalizations alone can be extraordinarily troublesome and debilitating. But these hundreds of patient histories going back almost two centuries, when combined with clinical experience, observations at TS support groups, and interactions with TS sufferers, reinforce the reality that the greatest distress and debility come with the most persistent and florid symptoms, especially those that include involuntary swearing. In part, this is because those with these symptoms suffer from the greatest public and private humiliation and thus, the most social isolation. By DSM-IV criteria, those who curse represent a statistical minority. In addition, recent studies have suggested that mild TS may be a common developmental disorder in young children. Weary of the association of TS as a cursing disease, patient support groups increasingly have attempted to change the public's perception of TS. Typical Tourette's has thus been shorn of its most well-known symptom.

In an attempt to more accurately construct a phenotype, we run the risk of sanitizing Tourette syndrome and unintentionally characterizing its most florid victims as "other" and "exceptions." In short, we run the risk of stigmatizing those who, through no fault of their own, are compelled by a brain malfunction to curse because they constitute a minority and no longer represent the entire problem. Yet, without the stories of these cursing ticcers, from the Marquise de Dampierre in the 1820s to Adam Seligman in the 1980s, the world never would have been educated about Tourette syndrome in the first place. Without the stories of florid sufferers, psychiatrists, neurologists, and researchers never would have been authorized (or funded) to investigate this disorder, and they never would have been able to treat the less florid majority. Are we now about to resegregate those with the most severe symptoms of TS as "abnormal" cases? Of course, the issue may not be put so starkly in the medical literature; rather, it will be addressed as "typing" or classification. However, having often witnessed parents of those with "mild" symptoms attempting to avoid contact with children with florid symptoms, I can imagine a scenario in which support groups segregate the florid from the simply ticcing and grunting. Such a result would be a gothic reenactment of the histories of attitudes toward Tourette syndrome sufferers elaborated in this book. With that in mind this book is a tribute to those who suffered the humiliating experience of cursing contrary to their desire not to do so. It is also a reminder to the rest of us of the debt we owe them.

Glossary

Ascertainment bias Conclusions made about an entire group based on a distorted or nontypical sample.

Agoraphobia An anxiety disorder characterized by abnormal fear of being in open, crowded, or public places.

Antrum cardicum The small passage where the esophagus joins the stomach.

Aphasia A brain injury or stroke. In Broca's aphasia, articulation of speech and language is affected; in Wernicke's aphasia, interpretation of speech is affected.

Arithmania An obsessive or compulsive need to count objects.

Assay A quantitative and qualitative analysis of a substance.

Athetosis A congenital disease in which those afflicted make slow, coarse, and writhing involuntary movements.

Autonomic nervous system That part of the central nervous system that regulates the vital functions of the body without any conscious control, including activity of the heart muscle, digestive system, and breathing.

Basal ganglia Part of the brain that regulates motor movements. It includes the caudate nucleus, the putamen, and the globus pallidus, which together are sometimes referred to as the corpus striatum (or striate cortex). Historically, including in most discussions of Tourette's, the basal ganglia and the corpus striatum have been used interchangeably.

Broca's area Part of the brain, located in the left temporal hemisphere in most right-handers and in the right temporal hemisphere of many left-handers, associated with the articulation of speech.

Caudate nucleus One of the structures that makes up the basal ganglia. Along with the globus pallidus and the putamen, the caudate forms the corpus striatum or striate cortex.

Clonic Involving the involuntary activity of the nerves.

Comorbid Independent illnesses or conditions manifested in a single patient.

Convulsive tic A synonym for the symptoms described by Gilles de la Tourette and labeled "la maladie des tics de Gilles de la Tourette."

Coprolalia Involuntary blurting out of curse words.

Copropraxia The acting out of explicitly sexual gestures or displays.

Corpus striatum (or striate cortex). A combination of the caudate nucleus, putamen, and globus pallidus. Also see basal ganglia.

Cross-reaction An antibody action in which a protein on one's own cells is mistaken for one on an invading bacteria or virus. Sometimes referred to as molecular mimicry.

Dopamine A neurotransmitter whose action is connected to a variety of motor and affective functions. The neurotransmitter involved in the brain circuits of the basal ganglia associated with the involuntary motor movements and tics in Tourette syndrome.

DSM (Diagnostic and Statistical Manual of Mental Disorders), published by the American Psychiatric Association; the main reference used by American (and other) psychiatrists.

Dystonia Any impairment of muscle tone. A condition commonly involving the head, neck, and tongue, and often occurring as an adverse effect of medication.

Echolalia Involuntary repetition of one's own or others' words or phrases.

Echokinesia A compulsion to imitate others' movements.

Echopraxia Imitation of others' behaviors or actions.

Enuresis Bed-wetting.

Epitope One's own cell protein mistaken by antibodies as a protein on invading bacteria or viruses.

Ethmoid sinus Sinus located near the ethmoid bone of the nasal area.

Etiology Origin or cause of a disease.

Globus pallidus One of the structures that makes up the basal ganglia.

Haloperidol (Haldol) An antipsychotic drug widely used to control motor tics and other symptoms of Tourette syndrome. Haloperidol inhibits dopamine transmission in the part of the brain that regulates the movements most associated with tics, the substantia nigra to the basal ganglia circuit.

Histology Study of cells.

Hypoplasia Underdevelopment of any part of the body, organ, or cell cluster.

Hypothalamus Brain structure below the thalamus that regulates the autonomic nervous system, endocrine system, and other body functions, including sleep, appetite, and temperature.

Idiopathic Of unknown origin.

Lesion Alteration of tissue through injury, accident, wound, or any other pathological cause.

Leucotomy Lobotomy.

Libido Psychoanalytic term for psychic energy associated with biological or instinctual drives.

Molecular mimicry An antibody action in which a protein (molecule) on one's own cells is mistaken for one on an invading bacteria or virus. Sometimes referred to as cross-reaction.

Myoclonus A spasm of a muscle or a group of muscles.

Neurasthenia A combination of fatigue, nervousness, and psychological symptoms attributed to exhaustion of a person's nervous system. The condition, first described by the American neurologist George M. Beard in the 1880s, was a popular diagnosis, especially in the United States, for the next two decades.

Neuroleptic Any drug that causes an altered state of consciousness (neurolepsis); an antipsychotic agent.

Neurotransmitter A brain chemical, such as serotonin, dopamine, or norepinephrine, that modifies or results in transmission of impulses between neurons.

Pathogenesis The underlying cause (etiology) of an illness or condition.

Phenotype Classification of a group or organism by a set of physical, behavioral, and psychological traits.

Putamen One of the brain structures of the basal ganglia. Along with the caudate nucleus and the globus pallidus, the putamen makes up the corpus striatum or striate cortex.

R.1625 Experimental name for haloperidol during its initial clinical trials in Europe and North America.

Scopolamine A sedative with various applications. Also known as hyoscine.

St. Vitus's dance Early term for Sydenham's chorea; in French, danse de St. Guy.

Striate cortex See corpus striatum.

Substantia nigra A cellular structure at the base of the brain responsible for the production of the neurotransmitter dopamine. Cell die-off in the substantia nigra, resulting in reduction of production of dopamine, is associated with Parkinson's disease.

Substrate Underlying substance acted on or altered that results in a condition that enables a symptom or set of symptoms.

Sydenham's chorea Ceaseless occurrence of a wide variety of rapid, jerky, involuntary but well-coordinated movements. Onset is usually between the ages of 5 to 15. Most patients recover in 2 to 3 months, with recurrence in about one-third of all cases. Symptom presentation follows a prior infection with Group A β hemolytic streptococcus (GABHS).

Tardive dyskinesia Involuntary repetitious facial, limb, and body tremors resulting from extended drug treatment, especially tranquilizers containing phenothiazine derivatives.

Thalamus Brain structure through which a wide variety of feelings and sensations are relayed to cortical areas of the brain, including pain, temperature, touch, feelings associated with pleasantness, unpleasantness, and a variety of arousal mechanisms.

Ticcers Those afflicted with uncontrollable motor movements (tics) and involuntary vocalizations; often used synonymously for what now is called Tourette syndrome; alternatively spelled "tiqueurs" or "tickers."

Tics Involuntary motor movements of the head, neck, limbs, and torso, including eye blinking, tongue protrusions, and shoulder shrugs. Vocal tics may include involuntary barks, grunts, yelps, and coughs.

Tiqueurs See ticcers.

Torticellis A (most often) congenital condition causing contraction of neck muscles that results in inclining the head to one side.

Transorbital lobotomy A procedure in which an ice pick-like implement was inserted through the eye sockets into the frontal brain lobes. In theory this procedure aimed to cut the connections between prefrontal lobes and the deeper structures connected with emotion, particularly the thalamus.

Notes

Preface

1. Howard I. Kushner, *Self-Destruction in the Promised Land: A Psychocultural Biology of American Suicide* (New Brunswick, N.J.: Rutgers University Press, 1989), pp. 114–117.

2. Alfred S. Evans, *Causation and Disease: A Chronological Journey* (New York: Plenum Publishing Corp., 1993), p. 2.

3. Joseph J. Hallett and Louise S. Kiessling, "Neuroimmunology of Tics and other Childhood Hyperkinesias," *Neurologic Clinics of North America*, 15 (1997): 333–344.

4. Howard I. Kushner and Louise S. Kiessling, "The Controversy over the Classification of Gilles de la Tourette's Syndrome, 1800–1995," *Perspectives in Biology and Medicine*, 39 (1996): 409–435.

5. Susan E. Swedo, Henrietta Leonard, Barbara B. Mittleman, Albert J. Allen et al., "Identification of Children with Pediatric Autoimmune Neuropsychiatric Disorders Associated with Streptococcal Infections by a Marker Associated with Rheumatic Fever," *American Journal of Psychiatry*, 154 (1997): 110–112.

6. This is already being done on an experimental basis. See A. J. Allen, H. L. Leonard, and S. E. Swedo, "Case Study: A New Infection Triggered, Autoimmune Subtype of Pediatric OCD and Tourette's Syndrome," *Journal of the American Academy of Child and Adolescent Psychiatry*, 34 (1995): 307–311.

1. An Elusive Syndrome

1. The diagnostic criteria are: (1) Multiple motor and one or more vocal tics must have been present for some time; (2) The tics must occur many times during the day (in bouts or intermittently) for more than a year—never tic free for more than a month; (3) The disturbance causes distress or significant impairment of functioning; (4) Onset before 18 years;

(5) Not due to direct physiological effects of medication or a general medical condition, Huntington's disease or post-viral encephalitis. *Diagnostic and Statistical Manual of Mental Disorders*, 4th ed., rev. ed. (Washington, D.C.: American Psychiatric Association, 1994), pp. 101–103.

2. For an informative overview, see Ruth Dowling Bruun and Bertel Bruun, *A Mind of Its Own: Tourette's Syndrome: A Story and a Guide* (New York: Oxford University Press, 1994).

3. For accounts written by those diagnosed with Tourette syndrome see Adam Ward Seligman and John S. Hilkevich, eds., *Don't Think about Monkeys* (Duarte, Calif.: Hope Press, 1992).

4. Caroline M. Tanner and S. M. Goldman, "Epidemiology of Tourette Syndrome," *Neurology Clinics,* 15 (1997): 395–402; Caroline M. Tanner, "Epidemiology," in Roger Kurlan, ed., *Handbook of Tourette's Syndrome and Related Tic and Behavioral Disorders* (New York: Marcel Dekker, 1993), pp. 337–344, see esp. pp. 339–341; Theodore Fallon, Jr. and Mary Schwab-Stone, "Methodology of Epidemiological Studies of Tic Disorders and Comorbid Psychopathology," in Thomas J. Chase, Arnold J. Friedhoff, and Donald J. Cohen, *Tourette Syndrome: Genetics, Neurobiology, and Treatment* (New York: Raven Press, 1992).

5. Anne Mason, Sube Banerjee, Valsamma Eapen, Harry Zeitlin, and Mary Robertson, "The Prevalence of Gilles de la Tourette's Syndrome in a Mainstream School Population," *Developmental Medicine and Child Neurology,* 40 (1998): 292–296. Mason et al. claim that "TS . . . is more common, milder and is associated with fewer other psychopathologies than has been suggested by previous studies." (p. 296); Also see Valsamma Eapen, Mary M. Robertson, Harry Zeitlin, and Roger Kurlan, "Gilles de la Tourette's Syndrome in Special Education Schools: A United Kingdom Study," *Journal of Neurology,* 244 (1997): 378–382.

6. David E. Comings, J. A. Himes, and Brenda J. Comings, "An Epidemiological Study of Tourette's Syndrome in a Single School District," *Journal of Clinical Psychiatry,* 51 (1990): 463–469.

7. Howard I. Kushner, *Self-Destruction in the Promised Land: A Psychocultural Biology of American Suicide* (New Brunswick, N.J.: Rutgers University Press, 1989), pp. 114–117.

8. Thus, diagnosis and treatment of syndromes are necessarily more tentative and more plastic than for diseases and have important therapeutic consequences. "Mistaking a symptom, sign, or marker for a disease," pharmacology professor Morely Sutter reminds us, "leads to confusion in treatment. Symptoms, signs, or markers can be altered without necessarily affecting the underlying disease process. Thus, drugs which directly reduce fever are unlikely to affect the outcome of a disease process which produces the fever." Morely C. Sutter, "Assigning Causation in Disease:

Beyond Koch's Postulates," *Perspectives in Biology and Medicine,* 39 (1996): 581–592, quotation from p. 585.

9. As Sutter points out, "signs and symptoms by themselves do not constitute a disease, but this distinction unfortunately is often ignored. To justify the label of disease the pathology should be able to be well described, and the stages and sequences in the development of the disease should be known and described. It is inappropriate to label as a disease a physical finding or sign a symptom, a phenomenon, or an observation." Thus "a fever is a finding or sign not a disease. Fever can have many causes ranging from excessive exposure to heat to malignancies and infections. Similarly hypertension (high blood pressure) is a finding or sign and not necessarily a disease." Ibid., p. 584.

10. Terra Ziporyn, *Nameless Diseases* (New Brunswick, N.J.: Rutgers University Press, 1992), pp. 1–2.

11. Alfred S. Evans, *Causation and Disease: A Chronological Journey* (New York: Plenum Publishing Corp., 1993), p. 2. Also see David E. Comings, "Tourette's Syndrome: A Behavioral Spectrum Disorder," in William J. Weiner and Anthony E. Land, eds., *Advances in Neurology, Behavioral Neurology of Movement Disorders* (New York: Raven Press, 1995), pp. 293–303; Mary M. Robertson, "The Gilles de la Tourette Syndrome: The Current Status," *British Journal of Psychiatry,* 154 (1989): 147–169.

12. Arthur K. Shapiro and Elaine S. Shapiro, "Evaluation of the Reported Association of Obsessive-Compulsive Symptoms or Disorder With Tourette's Disorder," *Comprehensive Psychiatry,* 33 (1992): 152–165.

13. David E. Comings, "Tourette Syndrome: A Hereditary Neuropsychiatric Spectrum Disorder," *Annals of Clinical Psychiatry,* 6 (1994): 235–247.

14. Not that there hasn't been a concerted effort. See the Tourette Syndrome Classification Study Group, "Definitions and Classifications of Tic Disorders," *Archives of Neurology,* 50 (1993): 1013–1016.

15. DSM-IV, pp. 101–103.

16. Donald Cohen to Sue Levi-Pearl (re: DSM-IV Criteria), 2 June 1994, Tourette Syndrome Association Archives, Bayside, L. I. Roger Freeman, Diane Fast, and M. Kent, "DSM-IV Criteria for Tourette's," *Journal of the American Academy of Child Adolescent Psychiatry,* 34 (1995): 400; David E. Comings, "Letter to the Editor, re: DSM-IV," ibid., p. 401; M. A. First, A. Frances, and H. Pincus, "DSM-IV Editors Reply," ibid., p. 402. Also, interview with Drs. Arthur K. and Elaine S. Shapiro, 21 September 1994, Scarsdale, New York.

17. James F. Leckman, Bradley S. Peterson, George M. Anderson, Amy F. T. Arnsten, David L. Pauls, and Donald J. Cohen, "Pathogenesis of Tourette's Syndrome," *Journal of Child Psychology and Psychiatry,* 38 (1997): 119–142, see discussion on pp. 122–123. For a similar view see

Roger Kurlan, "Future Direction of Research in Tourette Syndrome," *Neurologic Clinics,* 15 (1997): 451–456.

In 1997 Roger Freeman and his colleagues at the Neuropsychiatry Clinic at Children's Hospital in Vancouver, with the help of others in Canada and North Dakota, organized the Canadian-American Tourette Syndrome Database (CATS), a multi-site database one of whose aims is to identify the prevalence of symptom clusters. The study has attracted participant physicians worldwide, with most initially from Canada and the United States. CATS will have evaluated several thousand cases by the time this book appears in print. The survey already has made a major contribution to the identification of those symptoms and symptom clusters most frequently attached to and accompanying TS. Although these findings will aid clinicians and researchers, CATS is not designed to identify the actual causes or mechanisms underlying Tourette's.

18. Oliver Sacks, *The Man Who Mistook His Wife for a Hat* (New York: Harper and Row, 1987), pp. 92–101.

19. Oliver Sacks, "A Surgeon's Life," *The New Yorker* (March 16, 1992), pp. 85–94. Reprinted in Oliver Sacks, *An Anthropologist on Mars* (New York: Knopf, 1995).

20. Serge Lebovici, J.-F. Rabain, Tobie Nathan, R. Thomas, and M.-M. Duboz, "A Propos de la Maladie de Gilles de la Tourette," *Psychiatrie de L'Enfant,* 29 (1986): 5–59; J.-P. Guéguen, "Le Syndrome de Gilles de la Tourette (Maladie des Tics)," *Soins Psychiatrie,* 110/111 (1989–1990): 25–27.

21. Weston La Barre, "Obscenity: An Anthropological Appraisal," *Law and Contemporary Problems,* 20 (1955): 533–543, esp. pp. 533–537; David B. Morris, "The Neurobiology of the Obscene: Henry Miller and Tourette Syndrome," *Literature and Medicine,* 12 (1993): 194–214, esp. pp. 205–210.

22. Timothy Jay, "The Role of Obscene Speech in Psychology," *Interfaces: Linguistics, Psychology, and Health Therapeutics,* 12 (1985): 75–91.

23. David E. Comings and Brenda Comings, "Tourette Syndrome: Clinical and Psychological Aspects of 250 Cases," *American Journal of Human Genetics,* 37 (1985): 435–450; Colin Martindale, "Syntactic and Semantic Correlates of Verbal Tics in Gilles de la Tourette's Syndrome: A Quantitative Case Study," *Brain and Language,* 4 (1977): 231–247, esp. pp. 244–245; Bryan W. Robinson, "Limbic Influences on Human Speech," *Annals of the New York Academy of Sciences,* 280 (28 October 1976): 761–771, esp. pp. 765–767; M. R. Trimble and Mary R. Robertson, "The Psychopathology of Tics," in Chase, Friedhoff, and Cohen, eds., *Tourette Syndrome,* p. 420.

24. Seligman and Hilkevich, *Don't Think about Monkeys,* pp. 1–2.

25. Howard I. Kushner, "Medical Fictions: The Case of the Cursing Marquise

and the (Re)Construction of Gilles de la Tourette's Syndrome," *Bulletin of the History of Medicine,* 69 (1995): 224–254.

2. The Case of the Cursing Marquise

1. Her words in French were "merde and foutu cochon," which literally translate into "shit and filthy pig," but the more accurate colloquial meaning of "foutu cochon" is "fucking pig." Her obituaries included descriptions of her "notorious" cursing. Georges Gilles de la Tourette, "Étude sur une Affection Nerveuse Caractérisée par de l'Incoordination Motrice Accompagnée d'Écholalie et de Coprolalie (Jumping, Latah, Myriachit)," *Archives de Neurologie,* 9 (1885): 19–42, 158–200, see pp. 26, 180.

2. Lucien Malson, *Wolf Children* (followed by *The Wild Boy of Aveyron* by Jean Itard), trans. Peter Ayrton, Joan White, and Edmund Fawcett (New York: Monthly Review Press, 1972), pp. 91–178. See also Harlan Lane, *The Wild Boy of Aveyron* (Cambridge, Mass.: Harvard University Press, 1976).

3. Jean M. G. Itard, "Mémoire sur Quelques Fonctions Involontaires des Appareils de la Locomotion, de la Préhension et de la Voix," *Archives Générales de Médecine,* 8 (1825); 385–407, quotation from p. 403.

4. Gilles de la Tourette, "Étude sur une Affection Nerveuse," p. 21.

5. Jean-Martin Charcot, *Charcot the Clinician: The Tuesday Lessons: Excerpts from Nine Case Presentations on General Neurology Delivered at the Salpêtrière Hospital in 1887–1888,* trans. and commentary Christopher Goetz (New York: Raven Press, 1987), pp. 56, 60.

6. See Harold Stevens, "Gilles de la Tourette and his Syndrome by Serendipity," *American Journal of Psychiatry,* 128 (1971): 489–491; Arthur K. Shapiro et al., *Gilles de la Tourette Syndrome* (New York: Raven Press, 1978), pp. 15–18; Paul Guilly, "Gilles de la Tourette," in F. C. Rose and W. F. Bynum, eds., *Historical Aspects of the Neurosciences* (New York: Raven Press, 1982), pp. 397–415; A. J. Lees, "Georges Gilles de la Tourette: The Man and His Times," *Revue Neurologique,* 112 (1986): 808–816; Michel Dugas, "La Maladie des Tics: D'Itard aux Neuroleptiques," *Revue Neurologique,* 142 (1986): 817–823; Arnold J. Friedhoff and Thomas N. Chase, eds., *Gilles De la Tourette Syndrome* (New York: Raven Press, 1982); Mary M. Robertson, "The Gilles de la Tourette Syndrome: The Current Status," *British Journal of Psychiatry,* 154 (1989): 147–169; David E. Comings, *Tourette Syndrome and Human Behavior* (Duarte, Calif.: Hope Press, 1990); Paul Sandor, "Gilles de la Tourette Syndrome: The Current Status," *Journal of Psychosomatic Research,* 37 (1993): 211–226.

7. See Ruth Dowling Bruun and Bertel Bruun, *A Mind of Its Own: Tourette's Syndrome: A Story and a Guide* (New York: Oxford University Press,

1994), pp. 8–10; Peter G. Como, "Obsessive-Compulsive Disorder in Tourette's Syndrome," in W. J. Weiner and A. E. Lang, eds., *Behavioral Neurology of Movement Disorders* (New York: Raven Press, 1995), 281–291, see p. 281.

8. Stanley Finger, *Origins of Neuroscience: A History of Explorations into Brain Function* (New York: Oxford University Press, 1994), pp. 233–234.

9. C. G. Goetz and H. L. Klawans, "Gilles de la Tourette on Tourette's Syndrome," in Friedhoff and Chase, eds., *Gilles de la Tourette Syndrome*, pp. 1–16. An English translation of the cases discussed in the first part of Gilles de la Tourette's two-part essay was published in 1998. However, Gilles de la Tourette's commentaries on these cases and on the illness itself, which make up the majority of his article, still await translation. See Graeme Yorston and Nick Hindley, "Study of a Nervous Disorder Characterized by Motor Incoordination with Echolalia and Coprolalia: The Introduction and Case Studies of Gilles de la Tourette's 1885 Paper," *History of Psychiatry*, 9 (1998), 97–120.

10. For a detailed discussion of Itard's article see Howard I. Kushner, "Medical Fictions: The Case of the Cursing Marquise and the (Re)Construction of Gilles de la Tourette's Syndrome," *Bulletin of the History of Medicine*, 69 (1995): 224–254, esp. pp. 232–236.

11. The place of narrative in the history of medicine has been discussed in Arthur Kleinman, *The Illness Narratives: Suffering, Healing, and the Human Condition* (New York: Basic Books, 1988); Kathryn M. Hunter, *Doctors' Stories: The Narrative Structure of Medical Knowledge* (Princeton: Princeton University Press, 1991); Nancy M. P. King and Ann Folwell Stanford, "Doctors' Stories and True Stories: A Cautionary Reading," *Literature and Medicine*, 11 (1992): 185–199; David H. Flood and Rhonda L. Soricelli, "Development of the Physician's Narrative Voice in the Medical Case History," ibid., pp. 64–83; and Ellen Dwyer, "Stories of Epilepsy, 1880–1930," in C. Rosenberg and J. Golden, eds., *Framing Disease: Studies in Cultural History* (New Brunswick, N.J.: Rutgers University Press, 1992), pp. 248–272.

12. Itard, "Mémoire sur Quelques Fonctions Involontaires," p. 385.

13. Lane, *The Wild Boy of Aveyron*, pp. 73–76.

14. For a discussion of the origins and development of the moral treatment in France, see Gladys Swain, *Le Sujet de la Folie: Naissance de la Psychiatrie* (Toulouse: Privat, 1977), esp. pp. 61–77; Jan Goldstein, *Console and Classify: The French Psychiatric Profession in the Nineteenth Century* (New York: Cambridge University Press, 1987), pp. 65–66, 80–119.

15. Itard asserted that "in women, involuntary movements without delirium are more common, but at the same time less serious. A simple nervous irritation leads among them [women] to sympathetic disorders, which among men results from a profound cerebral lesion." Itard, "Mémoire sur Quelques Fonctions Involontaires," p. 396.

16. Howard I. Kushner, "Suicide, Gender, and the Fear of Modernity in Nineteenth-Century Social Thought," *Journal of Social History,* 26 (1993): 461–490.

17. Itard, "Mémoire sur Quelques Fonctions Involontaires," pp. 396–397, 400.

18. Ibid., pp. 398–399, 402–403.

19. Ibid., p. 404.

20. Ibid., p. 406.

21. "Alienist" was the nineteenth-century term for those physicians who treated a range of seeming mental disturbances. Given their general assumption of the relationship between the mental and the physical interactions of mental illness and insanity, to refer to these physicians as either psychiatrists or neurologists would be misleading.

22. Ernest Billod, "Maladies de la Volonté," *Annales Médico-Psychologiques,* 10:1, published in 3 parts, (1847): 15ff; 170–202; 317–347. For Billod's discussion of Dampierre see 3: 321–323.

23. See Germain E. Berrios and M. Gili, "Will and Its Disorders: a Conceptual View," *History of Psychiatry,* 6 (1995): 87–104, see pp. 89–90, 96–97. See also Edwin Clark and L. S. Jacyna, *Nineteenth-Century Origins of Neuroscientific Concepts* (Berkeley: University of California Press, 1987), pp. 269–270, see n232. Also see Anne Harrington, *Medicine, Mind and the Double Brain: A Study in Nineteenth-Century Thought* (Princeton: Princeton University Press, 1987), pp. 8–9, 38–39.

24. Billod, "Maladies de la Volonté," 3: 321–323; quotation from p. 336, italics added. See also Berrios and Gili, "Will and Its Disorders," p. 89.

25. Ibid., pp. 321–323.

26. David Didier Roth, *Histoire de la Musculation Irrésistible ou de la Chorée Anormale* (Paris: Ballière, 1850), pp. 196–206.

27. C. M. S. Sandras, *Traité Pratique des Maladies Nerveuses,* 2 vols. (Paris: Germer-Baillière, 1851), II: 530–534.

28. See Théodule Ribot, *Les Maladies de la Volonté,* 14th ed. (Paris: Félix Alcan, 1900). In what follows I quote from the English translation of the 8th French edition, which is unchanged from the 1883 edition. See *The Diseases of the Will,* trans. Merwin-Marie Snell, from the 8th French edition of 1894 (Chicago: The Open Court Publishing Company, 1915).

29. Ian R. Dowbiggin, *Inheriting Madness: Professionalization and Psychiatric Knowledge in Nineteenth-Century France* (Berkeley: University of California Press, 1991), p. 130; see also pp. 1–10; 116–143.

30. For more on degeneration theory see Robert A. Nye, *Crime, Madness, and Politics in Modern France: The Medical Concept of National Decline* (Princeton: Princeton University Press, 1984), pp. 121–131; 143–144; Ian Dowbiggin, "Degeneration and Hereditarianism in French Mental Science, 1840–90: Psychiatric Theory as Ideological Adaption," in W. F. Bynum, R. Porter, and M. Shepherd, eds., *The Anatomy of Madness:*

Essays in the History of Psychiatry, 2 vols. (New York: Tavistock Publications, 1985), 1: 188–232; Daniel Pick, *Faces of Degeneration: A European Disorder, c. 1848–c. 1918* (New York: Cambridge University Press, 1989); Mark S. Micale, *Approaching Hysteria: Disease and Its Interpretation* (Princeton: Princeton University Press, 1995), 203–220.

31. Ribot, *The Diseases of the Will,* p. 55.
32. Ibid., quotations from pp. 56–57, 66–67.
33. Gilles de la Tourette directed readers to see Ribot's introduction and second chapter. Gilles de la Tourette, "Étude sur une Affection Nerveuse," p. 200.
34. Ibid., p. 194.
35. Georges Gilles de la Tourette, "La Maladie des Tics Convulsifs," *La Semaine Médicale,* 19 (1899): 153–156; quotations from pp. 155–156; Gilles de la Tourette, "Étude sur une Affection Nerveuse," p. 188.
36. Shapiro et al., *Gilles de la Tourette Syndrome,* p. 18.
37. "En 1825, Itard publiait une observation qui était intégralement rapportée par Roth en 1850, et par Sandras en 1851. Cette observation que l'on trouvera en tête de celles que nous avons recueillies est extrêmement concluante et d'autant plus intéressante que la malade qui en fait l'objet a vécu jusqu'en 1884, et a été vue par M. le professeur Charcot, qui a contrôlé le diagnostic rétrospectif." Gilles de la Tourette, "Étude sur une Affection Nerveuse," p. 21. Goetz and Klawans inexplicably translate the French as "In 1825, Itard published a case report on a patient who was later reported by Roth in 1850 and by Sandras in 1851. The patient lived until 1884 and was finally seen and diagnosed by Professor Charcot." Goetz and Klawans, "Gilles de la Tourette on Tourette's Syndrome," p. 2.
38. Gilles de la Tourette, "Étude sur une Affection Nerveuse," p. 26, italics mine.
39. Jean-Martin Charcot, *Leçons du Mardi à la Salpêtrière Policliniques, 1887–1888,* p. 61, italics added. For a complete discussion, see Kushner, "Medical Fictions: The Case of the Cursing Marquise," pp. 237–241.
40. Jean-Martin Charcot, "Intorno ad alcuni casi di tic convulsivo con coprolalia ed ecolalia: Analogia col jumping di Beard, il latah della Malesia ed il miriachit di Hammond," lezione raccolta dal G. Melotti, *La Riforma Medica,* 1885, 1: 184: 2, 185: 2, 186: 2: part II, quotation from p. 2 (italics in original). Republished in J. M. Charcot, *Nuove Lezioni sulle Malattie del Sistema Nervoso,* raccolte dal Dr. Giulio Melotti (Milano: Antica Casa Editrice, Dottor Francesco Vallardi, 1887), pp. 153–165.
41. Gilles de la Tourette, "Étude sur une affection nerveuse," p. 26.
42. For the marquise's obituaries see ibid.
43. See Jacques Gasser, *Aux Origins du Cerveau Moderne: Localisations, Langage, et Mémoire dans l'Ouevre de Charcot* (Paris: Fayard, 1995), p. 33.

44. Stevens, "Gilles de la Tourette and His Syndrome"; Guilly, "Gilles de la Tourette," pp. 398–399; Lees, "Gilles de la Tourette," pp. 809–810; Goetz, *Charcot the Clinician,* p. 60.

45. George M. Beard, "Experiments with the 'Jumpers' or 'Jumping' Frenchmen of Maine," *Journal of Nervous and Mental Disorders,* 7 (1880): 478–490; George M. Beard, "Experiments with the 'Jumpers' of Maine," *Popular Science Monthly,* 18 (1880): 170–178. Beard also had presented a preliminary account in 1878. See George M. Beard, "Report to the American Neurological Association, 1878," *Journal of Nervous and Mental Disorders,* 5 (1878): 526.

46. Beard, "Experiments with the 'Jumpers' or 'Jumping' Frenchmen of Maine," p. 487.

47. Ibid., pp. 487–489.

48. Georges Gilles de la Tourette, trans. "Les 'Sauteurs' du Maine (Etats-Unis) par G. Beard," *Archives de Neurologie,* 5 (1881): 146–150; Gilles de la Tourette translated the essay that appeared in *Popular Science Monthly.*

49. Georges Gilles de la Tourette, "Jumping, Latah, Myriachit," *Archives de Neurologie,* 8 (1884): 68–84.

50. Ibid., p. 74 Gilles de la Tourette added that he had recently encountered four other cases, which "will be reported on his next work on this subject." For more on Gilles de la Tourette's 1884 article see Clara Lajonchere, Marsh Nortz, and Stanley Finger, "Gilles de la Tourette and the Discovery of Tourette's Syndrome: Includes a Translation of his 1884 Article," *Archives of Neurology,* 53 (1996): 567–574.

51. Gilles de la Tourette, "Étude sur une Affection Nerveuse." While some authors later claimed that Charcot immediately rejected the connection between latah and convulsive tics, the evidence seems convincing that, at least initially, Charcot fully endorsed the connection. For a detailed discussion see Kushner, "Medical Fictions: The Case of the Cursing Marquise," pp. 236–237, esp. n53.

52. Gilles de la Tourette, "Étude sur une Affection Nerveuse," pp. 194, 188. Also see Gilles de la Tourette, "La Maladie des Tics Convulsifs," pp. 155–156. Although Gilles de la Tourette noted that coprolalia often was absent, even in his observations, he insisted that "in the great majority of cases, and at some time during the course of the disease, it will almost certainly make its appearance." "Étude sur une Affection Nerveuse," p. 184.

53. Charcot, *Leçons du Mardi à la Salpêtrière Policliniques, 1887–1888,* 22 March 1886, pp. 86–87.

54. Gilles de la Tourette, "Étude sur une Affection Nerveuse," pp. 33–36, 39–40.

55. Ibid., pp. 26–30, 36–38. Two other examples briefly discussed were not examined by Gilles de la Tourette. The first, a female patient of

Dr. A. Pitres of Bordeaux, reportedly experienced the onset of her symptoms at age nine. These included involuntary movements of her limbs and face, a guttural sound, coprolalia, echolalia (in the form of imitating barking dogs), and echopraxia (pp. 40–42). The final case, a twenty-three-year-old man, described by Charles Féré, an adjunct at the Salpêtrière, first experienced symptoms at fourteen, including general muscular incoordination and coprolalia (pp. 158–161).

56. Thus, rarely, if ever, are Gilles de la Tourette's other examples cited, discussed, or analyzed. A partial exception is found in Dugas who compares Gilles de la Tourette's general conclusions about these cases with the categories of the DSM-III (1980) and concludes that six of Gilles de la Tourette's nine cases mesh with the DSM-III (1980). But only three of these were actually patients observed by Gilles de la Tourette and none demonstrate Gilles de la Tourette's central claim of a life-long affliction. The remaining three (including the famous marquise) were based on textual examples. See Michel Dugas, "La Maladie des Tics: D'Itard aux Neuroleptiques," *Revue Neurologique,* 11 (1986): 817–823.

3. A Disputed Illness

1. Georges Gilles de la Tourette, "Étude sur une Affection Nerveuse Caractérisée par de l'Incoordination Motrice Accompagnée d'Écholalie et de Coprolalie (Jumping, Latah, Myriachit)," *Archives de Neurologie,* 9 (1885): 19–42, 158–200, see pp. 184, 194.

2. Jean-Martin Charcot, *Leçons du Mardi à la Salpêtrière Policliniques, 1887–1888* (Notes de Cours de MM. Blin, Charcot, et Colin) handwritten and printed. (Paris: Bureaux du Progrès Médical, 1887), 22 March 1886, pp. 86–87; Georges Gilles de la Tourette, "La Maladie des Tics Convulsifs," *La Semaine Médicale,* 19 (1899): 153–156; Gilles de la Tourette, "Étude sur une Affection Nerveuse," p. 188.

3. Ibid., p. 200.

4. For a discussion of Magnan's views and career see Ian Dowbiggin, "Back to the Future: Valentin Magnan, French Psychiatry, and the Classification of Mental Diseases, 1885–1925," *Social History of Medicine,* 9 (1996), 383–408.

5. Quoted from Julien Noir, *Étude sur les Tics* (Thèse pour le Doctorat en Médecine), (Paris: Bureaux du Progrès Médical: Faculté de Médecine de Paris, 1893; quotation from pp. 7, 103. Also see Maurice Legrain, *Du Délire chez les Dégénérés: Observations Prises à l'Asile Sainte-Anne, 1885–1886* (Paris: Progrès médical, 1888, reprinted Nendeln: Kraus, 1978), 87–88; and M. Legrain, "Deux Observations de Coprolalie des Dégénérés," *Annales Médico-Psychologiques,* 2 (1889): 118–120.

6. Georges Guinon, "Sur la Maladie des Tics Convulsifs," *Revue de*

Médecine, 6 (1886): 50–80; "Tics Convulsifs et Hystérie," *Revue de Médecine,* 7 (1887): 509–519; "Tics Convulsifs," *Dictionnaire Encyclopédique des Sciences Médicales* (Paris: Asselin et Houzeau, 1887), 555–588.

7. Ibid., p. 519.

8. Guinon, "Sur la Maladie des Tics Convulsifs," pp. 70–74.

9. Although influenced by Charcot's theory of traumatic hysteria, in Freud's later theory of conversion hysteria, the earlier incident is not an actual event, but a repressed wish.

10. For detailed discussions of Charcot's construction of hysteria, especially as it related to heredity, see Christopher Goetz, Michael Bonduelle, and Toby Gelfand, *Charcot: Constructing Neurology* (New York: Oxford University Press, 1995), pp. 260–263, 179–216; Mark Micale, "Charcot and les Névroses Traumatiques: Historical and Scientific Reflections," *Revue Neurologique,* 150 (1994): 499–500; Mark S. Micale, *Approaching Hysteria: Disease and Its Interpretation* (Princeton: Princeton University Press, 1995), pp. 202–220, 87–107; Daniel Widlöcher and N. Dantchev, "Charcot et l'Hystérie," *Revue Neurologique,* 150 (1994): 490–497.

11. Jean-Martin Charcot, "Toux et Bruits Laryngés chez les Hystériques, les Choréiques, les Tiqueux and dans quelques autres Maladies des Centres Nerveux," *Leçons du Mardi,* 22 March 1886, republished in *Archives de Neurologie,* 23 (1892): 69–88, see pp. 86–87.

12. Charcot, *Leçons du Mardi, 1887–1888,* 4 June 1889, pp. 468–469. Sometimes Charcot stretched to interpret family behaviors as fitting his hereditarian assumptions. In the case of a young woman whom Charcot diagnosed as having Gilles de la Tourette's illness, he admitted that "heredity seemed to be lacking in the sense that neither her father nor her mother seemed to have had any nervous illnesses." But, in what surely was one of the most indirect claims of inherited tics, Charcot added that when her mother was pregnant with the patient, she had "been employed by a banker who had a tic. And, our patient says, the constant viewing of her [mother's] employer, caused a nervous illness, sufficient to distress her [mother's] pregnancy and give to the infant . . . the tic you have observed." Charcot, "Des Tics et des Tiqueurs," *La Tribune Médicale,* 19 (1888): 571–573, quotation from p. 573.

13. Charcot, "Toux et Bruits Laryngés," p. 84.

14. Ibid., p. 54.

15. Ibid., pp. 60–61, 69.

16. Guinon, "Tics Convulsifs et Hystérie," p. 519.

17. Ibid.

18. Charcot, *Leçons du Mardi, 1887–1888,* p. 60.

19. Ibid., 13 December 1887, p. 61.

20. Using two cases of boys with motor tics as his illustration, Charcot at-

tempted to demonstrate that rhythmic chorea, a hysterical disorder, was a sequel to an earlier fright or startle experience, while maladie des tics could always be traced to a family history of insanity. Ibid., 17 January 1888, pp. 129.

21. Ibid., 24 January 1888, pp. 152–153.
22. See ibid., 21 February 1888, pp. 209–213.
23. Jacques Catrou, *Étude sur la Maladie des Tics Convulsifs (Jumping— Latah—Myriachit)* (Thèse pour le Doctorat en Médecine) no. 129, (Paris: Faculté de Médecine de Paris, Henri Jouve, 1890).
24. Ibid., pp. 12–23. Catrou did not dispute Guinon's observation that most ticcers displayed obsessive behaviors but merely repeated Charcot's earlier statement, which Gilles de la Tourette subsequently endorsed, that "there are *idea tics* just as there are *motor tics*" (p. 45, italics in original).
25. Ibid., pp. 35–36.
26. Ibid., pp. 35–38, 40–45.
27. Noir, *Étude sur les Tics,* pp. 167, 14–15; also see pp. 16–30. This class of tics seemed most closely to resemble what Charcot had called rhythmic hysteria. Charcot, *Leçons du Mardi, 1887–1888,* 17 January 1888, pp. 129.
28. Noir, *Étude sur les Tics,* pp. 165, 31–62.
29. Ibid., pp. 165, 126, 103–106; also see pp. 143–155.
30. Ibid., pp. 166, 157–158; also see pp. 91–142.
31. Mark S. Micale, "On the 'Disappearance' of Hysteria: A Study in the Clinical Deconstruction of a Diagnosis, *Isis,* 84 (1993); 496–526, see p. 517. Also see Christopher G. Goetz, "The Salpêtrière in the Wake of Charcot's Death," *Archives of Neurology,* 45 (1988), 444–447; Harrington, *Medicine, Mind, and the Double Brain,* pp. 179–182; Anne Harrington, "Metals and Magnets in Medicine: Hysteria, Hypnosis, and Medical Culture in Fin de Siècle Paris," *Psychological Medicine,* 19 (1988): 245–249; Jean-Matin Charcot, *Clinical Lectures on Diseases of the Nervous System,* trans. Thomas Savill (London: The New Sydenham Society, 1889; reprint, with an introduction by Ruth Harris, New York: Routledge, 1991), p. ix.
32. Jean-Martin Charcot, *Charcot the Clinician: The Tuesday Lessons, Excerpts from Nine Case Presentations on General Neurology Delivered at the Salpêtrière Hospital in 1887–1888,* trans. and commentary Christopher Goetz (New York: Raven Press, 1987), p. 69.
33. Édouard Brissaud, "La Chorée Variable des Dégénérés," *Revue Neurologique,* 4 (1896): 417–431, esp. pp. 428–431. Also see Édouard Brissaud, "La Chorée Variable des Dégénérés," in Henry Meige, ed., *Leçons sur les Maladies Nerveuses,* 2d ed. (Paris: Masson et Companie, 1899), pp. 516–541.

34. Brissaud, "La Chorée Variable des Dégénérés" (1896), p. 419; see also p. 429.

35. Gilles de la Tourette, "La Maladie des Tics Convulsifs," pp. 153–156.

36. Addressing Guinon directly, Gilles de la Tourette admitted that "you will sometimes find certain cases difficult to diagnose," but maintained that Guinon could "extricate yourself from this unpleasantness only by the thorough understanding that you have acquired previously from the illness of convulsive tics." Ibid., pp. 153, 156.

37. Ibid., p. 156.

38. In the majority of cases the onset of Sydenham's chorea is diagnosed between the ages of five and fifteen, with females three times more likely than males to contract it. Most patients recover in two to three months, but recurrence takes place in about one-third of all cases, sometimes more than once. Roger Bannister, *Brain's Clinical Neurology,* 6th rev. ed. (Oxford: Oxford University Press, 1986), pp. 93, 348–349. Also see S. E. Swedo, "Sydenham's Chorea: A Model for Childhood Autoimmune Neuropsychiatric Disorders," *JAMA,* 272 (1994): 1788–1791.

39. Sydenham also labeled hysteria as "chorea major." The modern history of chorea began with a late fourteenth and early fifteenth century epidemic of convulsive-like behaviors in a number of German villages. Because, in search of relief, some of the afflicted made a pilgrimage to St. Vitus's Chapel in Dresselhausen (in the district of Ulm in Swabia), the behavior was labeled "St. Vitus's Dance."

40. Other choreas may be congenital, as in athetosis, where the movements are slower, coarser, and more writhing and the recovery more problematic.

41. Sometimes referred to as "rheumatism." See Peter C. English, "Emergence of Rheumatic Fever in the Nineteenth Century," in Charles E. Rosenberg and Janet Golden, eds., *Framing Disease: Studies in Cultural History* (New Brunswick, N.J.: Rutgers University Press, 1991), pp. 21–32. Also see Gene H. Stollerman, "Changing Streptococci and Prospects for the Global Eradication of Rheumatic Fever," *Perspectives in Biology and Medicine,* 40 (1997): 165–189.

42. Richard Bright, "Cases of Spasmodic Disease Accompanying Affections of the Pericardium," *Transaction of the Medico-Chirurgical Society of London,* 22 (1838): 1–19; quotation from pp. 10–11.

43. Etienne Michel Bouteille, *Traité de la Chorée, ou, Danse de St. Guy* (Paris: Vincard, 1810), cited by Armand Trousseau, *Lectures on Clinical Medicine,* trans. and ed. Sir John Rose Cormack & P. Victor Bazire, 3rd ed. (Philadelphia: Lindsay & Blakiston, 1873), p. 831.

44. Bright, "Cases of Spasmodic Disease," p. 10.

45. Trousseau, *Lectures on Clinical Medicine,* p. 831.

46. Germain Sée, "De la Chorée. Rapports de Rhumatisme et des Maladies du Coeur avec les Affections Nerveuses et Convulsives," (written in 1845), *Mémoire de l'Académie de Médecine,* 15 (1850): 373–525; quotation from p. 390. Also see pp. 414, 381.
47. Ibid., pp. 382–383.
48. Trousseau, *Lectures on Clinical Medicine,* p. 831.
49. Bright, "Cases of Spasmodic Disease," pp. 1–19.
50. Sée, "De la Chorée," pp. 374, 381–384; 414–420; also pp. 390–401.
51. C. M. S. Sandras, *Traité Pratique des Maladie Nerveuses,* 2 vols. (Paris: Germer-Baillière, 1851) II: 518–519, 522–523, 528.
52. Ibid., II: 509–534.
53. Ibid., II: 522, see also pp. 523, 528. Along with rheumatism, Sandras saw a protean set of possible etiologies for choreas, including masturbatory habits, diet, constitution, and brain lesions. Ibid., II: 526–529.
54. Ibid., II: 513, 516–517.
55. In this chapter I have cited the 1873 English language translation of Trousseau's third edition: *Lectures on Clinical Medicine,* pp. 853, 832, 843, 847.
56. Ibid., pp. 856, 830, italics in original.
57. Alain Lellouch, "Charcot, Découvreur de Maladies," *Revue Neurologique,* 150 (1994): 506–510; Christopher Goetz, "Charcot, Scientifique Bifrons (A Scientific Janus)" *Revue Neurologique,* 150 (1994): 485–489.
58. Jacques Gasser, "Jean-Martin Charcot (1825–1893) et le Système Nerveux: Étude de la Motricité, du Langage, de la Memoire et de l'Hystérie à la Fin du XIXè Siècle," Ph.D. diss., Ecole des Hautes Études en Sciences Sociales, 1990, 199–207. "Chorea," wrote Charcot, "has been considered by several authors to be an expression of rheumatic fever . . . because chorea often develops following acute rheumatic fever." However, insisted Charcot, "chorea can exist without having anything to do with rheumatism." Charcot, *Leçons Du Mardi 1887–1888,* 6 December 1887, p. 38.
59. Gasser, "Jean-Martin Charcot (1825–1893) et le Système Nerveux," p. 205; Charcot, *Leçons du Mardi, 1887–1888,* pp. 38–39.
60. Gilles de la Tourette, "Étude sur une Affection Nerveuse," p. 200.
61. Noir, *Étude sur les Tics,* p. 167.
62. Gasser, "Jean-Martin Charcot (1825–1893) et le Système Nerveux," p. 205; Charcot, *Leçons du Mardi, 1887–1888,* pp. 38–39.
63. Charcot, *Leçons du Mardi, 1887–1888,* p. 39.
64. Even some at the Salpêtrière like Brissaud saw infection as playing an etiological role in Sydenham's. Brissaud, "La chorée variable des dégénérés," p. 428. Also see p. 424 for a clinical example. See also Édouard Brissaud, "Chorée Variable," *La Presse Médicale,* 13 (1899): 73–74.

65. E. Cadet de Gassicourt, "Chorées Rhumatismales et autres 196 Observations Inédites par le Professeur G. Sée," *Médecine Moderne,* 2 (1891): 725–726, quotation, p. 725.

66. George A. Brown, "Chorea: Its Relation to Rheumatism and Treatment," *Montreal Medical Journal,* 19 (1890): 581–588; quotation from p. 586.

67. W. M. Donald, "The Relation between Chorea and Rheumatism," *Physician and Surgeon,* 14 (1892): 535–538; quotation from p. 536.

68. Vincent Pagliano, "Rhumatisme et Chorée," *Marseille Médicale,* 28 (1891): 329–334. For a similar analysis see J. W. Shaw, "Relation between Chorea and Rheumatism," *The Dominion Medical Monthly and Ontario Medical Journal,* 6 (1896): 145–147.

69. American Pediatric Society, "Relation of Rheumatism and Chorea," *Archives of Pediatrics,* 10 (1893): 1–27.

70. G. Tully Vaughan, "Case of Chorea, due Probably to Rheumatism and Endocarditis," *Virginia Medical Monthly,* 9 (1882): 669, italics in original.

71. He identified the bacterium in solution as a "staphylococcus," which given the primitive state of early Pasteurian identification processes, was as likely streptococcus. E. Leredde, "Note sur un Cas d'Endocardite Choréique d'Origine Microbienne Probable," *Revue Mensuelle des Maladies de l'Enfance,* 9 (1891): 217–220; quotation from p. 217. Also see P. Duroziez, "Chorée—Diagnostic des Lésions du Coeur," *L'Union Médicale: Journal des Interéts Scientifiques et Pratiques Moraux et Professionals du Corps Médical,* 3 (1892): 711–724, esp. p. 716.

72. Henry P. Cooke, "The Relation of Chorea to Rheumatism," *Transactions of the Texas State Medical Association,* (1892): 146–153, quotation from pp. 152–153.

73. Robert Massalongo, "Chorée chez Deux Cardiques," *Revue Neurologique,* 3 (1895): 610–615; quotation from p. 613, italics in original.

74. Gilles de la Tourette had emphasized "the absence of any anatomical pathology," by which he meant observable cortical lesions. "Étude sur une Affection Nerveuse," p. 200.

75. Ibid., p. 33.

76. Fulgence Raymond, "Maladie des Tics avec Coprolalie," *Revue de l'Hypnotisme et de la Psychologie Physiologique,* 16 (1901): 126.

77. An examination of the frequency of diagnoses of "dégénéréscence mentale" or "dégénéréscence stigmats" for patients admitted to the Salpêtriere from the mid-1880s to the 1920s provides persuasive evidence in this regard. Archives de la Salpêtrière, *Registres d'Observation Médicale,* 5ᵉ division 1ère à 4ᵉ section, 6 R 15 à 6 R 20, 1884–1910; 6 R 42 à 6 R 43, 1885–1922; 6 R 50 à 6 R 55, 1885–1910; 6 R 63 à 6 R 73, 1873–1919; *Registre de Diagnostics:* 6 R 90, 1880–1883, Assistance Publique, Hopitaux de Paris.

78. C. L. Dana and W. P. Wilkin, "On Convulsive Tic with Explosive Disturbances of Speech (So-Called Gilles de la Tourette's Disease)," *Journal of Nervous and Mental Disease*, 13 (1886): 407–412, quotations from p. 409.

79. Ibid., pp. 407, 409, p. 412.

80. R. R. Stawell, "Habit Spasm in Children," *The Australian Medical Journal*, 17 (1895): 153–158, quotation from p. 155.

81. J. C. Wilson, "A Case of Tic Convulsif," *Archives of Pediatrics*, 14 (1897): 881–887.

82. Ibid., pp. 881–885.

83. Ibid., pp. 881, 885–886.

84. As early as 1851, Sandras suggested a connection between masturbatory habits and movement disorders, but not until the 1890s would this become a widely accepted view. See C. M. S. Sandras, *Traité Pratique des Maladie Nerveuses*, 2 vols. (Paris: Germer-Baillière, 1851), II: 526–529.

85. For discussions of masturbation as a disease see Robert P. Neuman, "Masturbation, Madness, and the Modern Concepts of Childhood and Adolescence," *Journal of Social History*, 8 (1975): 1–27, and H. Tristram Engelhart, Jr., "The Disease of Masturbation: Values and the Concept of Disease," *Bulletin of the History of Medicine*, 48 (1974): 234–248.

86. Otto Lerch, "Convulsive Tics," *American Medicine* (2 November 1901), 694–695, quotation from p. 695.

87. Ibid., pp. 694–695.

88. Angelo Taranta and Gene H. Stollerman, "The Relationship of Sydenham's Chorea to Infection with Group A Streptococci," *American Journal of Medicine*, 20 (1956): 170–175.

89. Henry Meige and E. Feindel, *Les Tics et leur Traitement*, préface de E. Brissaud (Paris: chez Masson, 1902).

90. Henry Meige and E. Feindel, *Tics and their Treatment*, with a preface by Professor Brissaud, trans. and ed. S. A. K. Wilson (New York: William Wood and Co., 1907).

4. The Case of "O." and the Emergence of Psychoanalysis

1. Henry Meige and E. Feindel, *Tics and their Treatment*, with a preface by Professor Brissaud, [revised and updated version of *Les Tics et leur Traitement* (1902)], trans. and ed. S. A. K. Wilson (New York: William Wood and Co., 1907), p. 13.

2. Ibid., pp. vii–x.

3. Ibid., pp. vii, 220, 225.

4. Sandor Ferenczi, "Psycho-Analytical Observations on Tic," *International Journal of Psycho-Analysis*, 2 (1921): 1–30, see p. 10.

5. Meige and Feindel, *Tics and their Treatment*, p. 1.

6. Ibid., p. 1.
7. Henry Meige and E. Feindel, "L'État Mental des Tiqueurs," *Progrès Médical* (7 September 1900), 146–149; Meige and Feindel, "Sur la Curabilité des Tics," *Gazette des Hôpitaux* (20 June 1901), 673–677.
8. Henry Meige, "Étude sur Cértaines Névropathes Voyageurs: Le Juif-Errant à la Salpêtrière," *Nouvelle Iconographie de la Salpêtrière,* 6 (1893): pp. 191–204, 277–291, 333–358. Meige's text was reprinted and edited with a preface by the French psychoanalyst Lucien Israël, *Le Juif-Errant à la Salpêtrière* (Paris: Collection Grands Textes, Nouvelle Objet, 1993).
9. Meige, *Le Juif-Errant à la Salpêtrière* (1993 edition), pp. 22–23, 27, 109.
10. Henry Meige, "Tics Variables, Tics d'Attitude," *Société du Neurologie de Paris (Bulletins Officiels)* (4 July 1901): 249–250; quotation from p. 249.
11. Henry Meige, "La Genèse des Tics," *Journal de Neurologie* (5 June 1902): 201–206; quotation from p. 201.
12. Ibid., p. 201.
13. Ibid., pp. 1–2.
14. Ibid., p. 108.
15. Ibid., pp. 107, 109, 117.
16. Ibid., p. 117. The supposed underdevelopment of higher cortical functions had its origins in the mid-nineteenth-century comparative "craniology" or skull measurements. For more see Stephen Jay Gould, *The Mismeasure of Man* (New York: W. W. Norton, 1981), 73–112; Anne Harrington, *Medicine, Mind and the Double Brain* (Princeton: Princeton University Press, 1987), pp. 63–69.
17. Ibid., p. 101.
18. Ibid., pp. 2, 13.
19. Ibid., pp. 3–5.
20. Ibid., pp. 5–6.
21. Ibid., pp. 10–11.
22. Ibid., pp. 12–13.
23. Ibid., pp. 16–17.
24. Ibid., p. 17.
25. Ibid., pp. 17–19.
26. Oliver Sacks, "Tics," *New York Review* (29 January 1987): 37–41, quotation from p. 38. Sacks repeats this argument almost verbatim in "Neuropsychiatry and Tourette's," *Neurology and Psychiatry: A Meeting of the Minds,* ed. Jonathan Mueller (Basel: Karger, 1989), pp. 156–174, quotation from pp. 160–162.
27. Meige and Feindel, *Tics and their Treatment,* pp. 21–22, 328–330.
28. Ibid., pp. 319–324. Also see Henry Meige and E. Feindel, "Traitement des Tics," *Presse Médicale* (16 March 1900): 125–127; Édouard Brissaud and E. Feindel, "Sur le Traitement du Torticolis Mental et des Tics Simi-

laires," *Journal de Neurologie* (15 April 1899): 141–149; René Cruchet, "Le Tic Convulsif et son Traitement Gymnastique (Mèthode de Brissaud et Mèthode de Pitres)," Thèse pour le Doctorat en Médecine (Bordeaux: G. Gounouilhou, Imprimeur de la Faculté de Médecine, Bordeaux, 1902), pp. 96–97.

29. Meige and Feindel, *Tics and their Treatment,* pp. 325–328; Albert Pitres, *Journal de Médecine de Bordeaux* (9 July 1899): 331; Albert Pitres, "Tic Convulsif Généralisés, Traités et Guéris par la Gymnastique Respiratoire," *Mémoires et Bulletins de la Société de Médecine et de Chirurgie* (1900): 431–436; Cruchet, "Le Tic Convulsif et son Traitement Gymnastique," pp. 137–138.

30. Ibid., p. 138.

31. Meige and Feindel, *Tics and their Treatment,* p. 329. Albert Pitres, "Tics convulsifs généralisés traités et guéres par la gymnastique respiration," *Journal de Médecine de Bordeaux,* 17 (1901): 107; Cruchet, "Le Tic Convulsif et son Traitement Gymnastique," p. 142.

32. Meige and Feindel, *Tics and their Treatment,* pp. 329, 335–340.

33. Ibid., pp. 229–230, 22.

34. 12th Congrès des Médecins, Aliénistes, et Neurologists de France and des Pays de Langue Française, Grenoble, 1–7 August 1902, in *Revue Neurologique,* 31 (1902): 766–798.

35. Ibid., pp. 773–778.

36. Ibid., p. 778, italics in original.

37. Joseph Grasset, "12th Congrès des Médecins, Aliénistes et Neurologists de France," pp. 782–784, italics in original.

38. Cruchet, "Le Tic Convulsif et son Traitement Gymnastique," p. 167.

39. Ibid., pp. 425–426.

40. René Cruchet, "Sur un Cas de Maladie des Tics Convulsifs," *Archives Générales de Médecine,* 83:1 (1906): 1180–1196.

41. Ibid., pp. 1182–1183, italics in original.

42. Ibid., pp. 1183–1189, 1192–1194, 1195–1196, italics in original.

43. Meige and Feindel, *Tics and their Treatment,* pp. 347–348.

44. Ibid., pp. 348–349.

45. Citing Meige and Feindel, a British physician wrote in 1907 that "the mental state of [ticcing] patients is frequently not quite normal, the will may be feeble, the power of concentration lacking, and the character in other ways unstable." Eric D. Macnamara, "Habit Spasm," *Westminster Hospital Reports* (1907): 48–58; quotation from p. 53.

46. E. W. Scripture, "Tics and their Treatment," *Archives of Pediatrics,* 26 (1909): 10–13; quotations from pp. 10–11.

47. A San Francisco physician explained in 1911 that childhood tics developed because "the fatigued mind is stimulated at a time when it should be rested . . . The treatment of tic" was "definitely and intimately connected with the etiology and symptomology (sic)" of neurasthenia and

"its prevention" was "inherently a part of the prophylaxis of all pediatric nervous states." E. C. Fleischner, "The Treatment of Tic in Childhood," 41st Annual Meeting of the California State Medical Society, *California State Journal of Medicine*, 9 (1911): 379–382; quotation from p. 379. For a discussion of neurasthenia see Barbara Sicherman, "The Use of a Diagnosis: Doctors, Patients, and Neurasthenia," *Journal of the History of Medicine*, 32 (1977): 33–54.

48. At the 1911 Meeting of the California Medical Society, Dr. G. V. Hamilton explained that "a tic begins as a reaction to a situation to which we cannot make a definite and satisfactory mental adjustment, and, according to the psychological mechanisms now so well described by Freud, persists automatically." "The Treatment of Tic in Childhood," 41st Annual Meeting of the California State Medical Society, *California State Journal of Medicine*, 9 (1911): 379–382; quotation from p. 382.

49. For a discussion of Ferenczi's relationship with Freud see Peter Gay, *Freud: A Life for Our Time* (New York: Norton, 1988), pp. 187–189, 576–587; Paul Roazen, *Freud and His Followers* (New York: Knopf, 1974), pp. 355–371.

50. Sandor Ferenczi, "Psycho-Analytical Observation on Tic," *International Journal of Psycho-Analysis*, 2 (1921): 1–30, quotation from p. 10.

51. Ibid., p. 7.

52. Sigmund Freud, *Studies in Hysteria* (1893–1895), "Frau Emmy von N," *The Standard Edition of the Complete Psychological Works of Sigmund Freud*, trans. and ed. by James Strachey, 23 vols. (London: The Hogarth Press, 1966), 2: 48–105; 307–309 (hereafter cited as SE).

53. See SE 2: 307–309 for a discussion of the chronology of Freud's treatment.

54. In Freud's later theory of conversion hysteria, the earlier incident is not an actual event, but a repressed wish.

55. SE 2: 49.

56. Ibid., pp. 77–83.

57. Ibid., pp. 77, 74–75, n2.

58. Ibid., p. 83. Richard Webster suspects that "no cure was effected in the case of Frau Emmy at any stage of Freud's treatment" and that "by taking advantage of a slight amelioration of her symptoms which was probably spontaneous, Freud was able to represent a negative outcome as a partial therapeutic success." Richard Webster, *Why Freud Was Wrong: Sin, Science, and Psychoanalysis* (New York: Basic Books, 1995), pp. 144, 148.

59. SE 2: 83–84. Rereading the case in 1924 Freud wrote: "I am aware that no analysts can read this case history today without a smile of pity. But it should be borne in mind that this was the first case in which I employed the cathartic procedure to a large extent." Ibid., p. 105, n1. A number of observers recently have taken Freud to task for failing to have diagnosed Frau Emmy's ticcing and vocalizations as Tourette syndrome. For a discussion of these criticisms see Howard I. Kushner, "Freud and the

Diagnosis of Gilles de la Tourette's Illness," *History of Psychiatry,* 9 (1998): 1–25.

60. Freud, "A Case of Hypnotic Treatment, *SE* 1: 116–128; quotation from p. 127, italics in original.

61. "No doubt," Freud continued, "only a small proportion of the involuntary movements occurring in *tics* originate this way. On the other hand, it would be tempting to attribute to this mechanism the origin of coprolalia." Freud, "A Case of Hypnotic Treatment," *SE* 1:127–128, italics in original.

62. Recently, the distinguished French psychiatrist Cyrille Koupernik suggested that Daniel Paul Schreber, whom Freud diagnosed in 1911 as a paranoid schizophrenic, may have suffered from Gilles de la Tourette's disease. See Cyrille Koupernik, "Schreber, Maladie de Gilles de la Tourette et Transsexualisme: A propos de 'L'esprit Assassiné' de M. Schatzman," *L'Evolution psychiatrique,* 48 (1983): 1065–1078. For an evaluation of this argument, see Kushner, "Freud and Gilles de la Tourette's Illness."

63. "The circumstance of the Tic being peculiar among neurotic phenomena gives strong support to the idea of Freud regarding the heterogeneous (organic) nature of the symptom." Ferenczi, "Psycho-Analytical Observations on Tic," pp. 2–3.

64. Ibid., pp. 4–5, 10, italics in original.

65. Throughout, Ferenczi spelled "coprolalia" as "copralalia."

66. Ibid., p. 1.

67. J. C. Wilson, "A Case of Tic Convulsif," *Archives of Pediatrics,* 14 (1897): 881–887, see p. 885; Otto Lerch, "Convulsive Tics," *American Medicine* (2 November 1901): 694–695, see p. 695.

68. Ferenczi, "Psycho-Analytical Observations on Tic," pp. 5–10.

69. Ibid., p. 6.

70. Ibid., pp. 5–6.

71. Ibid., p. 19.

72. Ibid., p. 22.

73. Oddly enough Ferenczi failed to make the connection, which Freud had made earlier, that sufferers of strokes in the language area of the brain (Broca's aphasias) and those with tics sometimes displayed involuntary outbursts of cursing. Perhaps this is what Freud had meant by his off-hand remark about the physiological substrate of tics.

74. Ibid., pp. 9–10, 22, quoting Meige and Feindel, *Les Tics et leur Traitement,* pp. 88–89.

75. Ferenczi, "Psycho-Analytical Observations on Tic," p. 23. Having appropriated Meige and Feindel's examples, Ferenczi was forgiving, noting that when Meige and Feindel wrote, "the doctrine of the unconscious was yet in its infancy" (p. 24).

76. Ibid., pp. 27–28.

77. There were a few exceptions. In 1925 Melanie Klein, one of the founders

of the "object relations" approach to psychoanalysis, wrote that she was in "essential disagreement" with Ferenczi's claim that "the tic is a primary narcissistic symptom having a common source with the narcissistic psychoses." Although Klein endorsed the view "that tic is an equivalent of masturbation," she was convinced that "the tic is not accessible to therapeutic influence so long as the analysis has not succeeded in uncovering the object relations on which it is based." Melanie Klein, "Zur Genese des Tics," *Internationale Zeitschrift Für Psychoanalyse,* 11 (1925): 332–349; reprinted and translated as "Contribution to the Psychogenesis of Tics," in Melanie Klein, *Contributions to Psychoanalysis, 1921–1945* (London: Hogarth Press, 1968), pp. 117–139; quotation from p. 133.

5. Competing Claims

1. Charles Trepsat, "Traitement d'un Tiqueur par la Psychanalyse," *Le Progrès Médical,* 35 (1922): 182–184; quotation from p. 182.
2. Ibid., p. 182.
3. Ibid., p. 184.
4. Raymond de Saussure, *La Méthode Psychanalytique,* with a preface by Sigmund Freud (Geneva: Payot, 1922). A translation of Freud's preface appears in Sigmund Freud, "Preface to Raymond de Saussure's *The Psycho-Analytic Method* (1922)," SE 19: 283–284.
5. Leon Chertok and Raymond de Saussure, *The Therapeutic Revolution, from Mesmer to Freud,* trans. R. H. Ahrenfeldt (New York: Brunner/Mazel, 1979), pp. v–vi.
6. Encephalitis is literally any infection of the central nervous system characterized by inflammation of the brain. Encephalitis may result either from neurotropic-virus, transmitted to humans from animal hosts via insect vectors, or as a sequel to a more general virus, or as a complication of mumps, measles, yellow fever or influenza. Less frequently, encephalitis can result from attenuated virus vaccines (smallpox and yellow fever) or from infection from the parasite sporozoa toxoplasma, found in birds and mammals, including humans. It also can result from infection of the protozoan trypanosoma cruzi, transmitted by the bite of the tsetse fly.
7. The infective agent was swine flu. For a discussion of the 1918–1926 pandemic see R. T. Ravenholt, "Encephalitis Lethargica," in Kenneth F. Kiple, ed., *The Cambridge World History of Human Disease* (New York: Cambridge University Press, 1993), pp. 708–712.
8. Raymond de Saussure, "Discussion sur l'Étiologie d'un Tic survenu Quinze Mois après une Encéphalite Léthargique Atypique," *Schweizer Archiv für Neurologie und Psychiatrie,* 12 (1923): 294–317.
9. Blanche was not told that she was adopted until she was ten years old. Blanche visited her natural mother only twice and "did not like her very much." Ibid., p. 298.

10. Ibid., p. 299.
11. Ibid., p. 305.
12. Ibid.; quotations from pp. 312, 314.
13. Freud, "Preface to Raymond de Saussure's *The Psycho-Analytic Method* (1922)," pp. 283–284.
14. Erwin Straus, "Über die organische Natur der Tics und der Koprolalie," *Zentralblatt für die gesamte Neurologie und Psychiatrie,* 47 (1927): 698–699, quotation from p. 698. Also see Erwin Straus, "Untersuchungen über die postchoreatischen Motilitätsstörungen, insbesondere die Beziehungen der Chorea minor zum Tic," *Monatsschrift für Psychiatrie und Neurologie,* 66 (1927): 261–324.
15. Straus, "Über die organische Natur der Tics und der Koprolalie," pp. 698–699; Straus, "Untersuchungen über die postchoreatischen Motilitätsstörungen," pp. 301–312. Straus rejected claims that convulsive tics had any psychogenic etiology. Ibid., pp. 264–265.
16. Ibid., pp. 268–271.
17. Straus, "Über die organische Natur der Tics und der Koprolalie," pp. 698–699.
18. Ibid., pp. 698–699.
19. He is best remembered for his identification of "Wilson's disease," a degenerative condition caused by malfunctioning in the body's processing of copper.
20. Henry Meige and E. Feindel, *Tics and their Treatment,* with a preface by Professor Brissaud [revised and updated version of *Les tics et leur Traitement* (1902)], trans. and ed. by S. A. K. Wilson (New York: William Wood and Co., 1907).
21. S. A. Kinnier Wilson, "The Tics and Allied Conditions," *The Journal of Neurology and Psychopathology,* 8 (1927): 93–109.
22. Ibid., p. 99.
23. Wilson found O.'s case to be "depressing therapeutically as it is entertaining clinically." Ibid., p. 101.
24. Ibid, pp. 96–98.
25. Ibid., p. 99.
26. Ibid., p. 100, italics in original.
27. Ibid., pp. 101–103.
28. René Cruchet, "The Tics and Allied Conditions," *The Journal of Neurology and Psychopathology,* 8 (1927): 104–106.
29. Ibid., p. 106.
30. Wilson, "Tics and Allied Conditions," pp. 106, 108. Cruchet had first written about this connection in 1917. R. Cruchet, Mousier, and Calmettes, "Quarante Cas d'Encéphalomyélite Subaiguë," *Société Médicales Bordelais* (27 April 1917); René Cruchet, "L'Encéphalomyélite à Bordeaux et dans les Régions Voisines," *Société de Médecine et de Chirurgie de Bordeaux,* 10 December 1920.

31. Cruchet however believed that convulsive tics could be "fairly easily distinguished" from those resulting from encephalitis. Cruchet, "Tics and Allied Conditions," p. 107.

32. Ibid., p. 107.

33. W. Russell Brain, "The Treatment of Tic (Habit Spasm)," *Lancet* (23 June 1928): 1295–1296; quotation from p. 1295.

34. Ibid., pp. 1295–1296.

35. Paul B. Beeson, "Fashions in Pathogenetic Concepts during the Present Century: Autointoxication, Focal Infection, Psychosomatic Disease, and Autoimmunity," *Perspectives in Biology and Medicine*, 36 (1992): 13–23, see esp. pp. 15–17. Also see R. A. Hughes, "Focal Infection Revisited," *British Journal of Rheumatology*, 34 (1994): 370–377.

36. Frank Billings, "Chronic Focal Infections and their Etiologic Relations to Arthritis and Nephritis," *Archives of Internal Medicine*, 9 (1912): 484–498; Frank Billings, *Focal Infection: The Lane Medical Lectures* (New York, Appleton, 1917).

37. E. C. Rosenow, "The Newer Bacteriology of Various Infections as Determined by Special Methods," *JAMA*, 63 (1914): 903–970 as cited in Beeson, "Fashions in Pathogenetic Concepts during the Present Century," p. 16.

38. For an excellent discussion of Rosenow's experiments see Hughes, "Focal Infection Revisited," pp. 373–375.

39. See J. W. Fairley, "Patrick Watson-Williams and the Concept of Focal Sepsis in the Sinuses: An Historical Caveat for Functional Endoscopic Sinus Surgery," *Journal of Laryngology and Otology*, 105 (1991): 1–6, esp. p. 4.

40. Henry Cotton, "The Relation of Chronic Sepsis to the So-Called Functional Mental Disorders," *Journal of Mental Science*, 69 (1923): 434–465.

41. Ibid., pp. 438–442, 460.

42. Archives, New York Hospital—Cornell Medical Center, New York, New York, hereafter cited as New York Hospital/Cornell Archives. For instance, in December 1922, an eleven-year-old female, diagnosed with chorea and habit spasm, had her tonsils and adenoids removed even though a physical exam revealed no infection. Case 24–5574, 2–15 December 1922. A sixteen-year-old boy, also diagnosed with "habit spasm," was admitted to New York Hospital in January 1924. His doctors decided to remove the patient's tonsils, only to learn that both tonsils and adenoids had been removed eight years earlier. Case 25–2684, 13–18 January 1924.

43. Parents often were less persuaded than physicians. For instance, a twelve-year-old girl's purposeless movements, twitching of the head, irritability, and coprolalia were not abated after a tonsillectomy. While her doctors considered whether to operate further, her mother took her home "against advice." Case 274, 22–27 August 1924, New York Hospital/Cornell Archives. In another typical case, a ten-year-old boy with uncoordinated leg

and head movements and vocalizations was found to have enlarged tonsils, but his parents "took child home after 2 days against Dr.'s advice," apparently unpersuaded that a tonsillectomy would cure their son's tics. Case 29–2304, 11–13 April 1930, ibid. The procedure was not without risk. For instance, a sixteen-year-old girl whose nervousness and twitchings had worsened died of what her doctors claimed was a "possible . . . pulmonary embolus" during a tonsillectomy. However, it seems more likely that she had a post-operative hemorrhage. Case 267866, July 1926, ibid. These procedures had become so routine that physicians at major hospitals felt that they had to justify *not* performing them. See Case 21–237, 3–24 January 1924, ibid.

44. Laurence Selling, "The Role of Infection in the Etiology of Tics," *Archives of Neurology and Psychiatry,* 22 (1929): 1163–1171; quotation from p. 1163.

45. Ibid., pp. 1163–1164.

46. Ibid., p. 1164.

47. For a description and illustrations of these procedures, see Fairley, "Patrick Watson-Williams and the Concept of Focal Sepsis in the Sinuses," pp. 2–5.

48. The antrum cardicum is the small passage where the esophagus joins the stomach.

49. Ibid., pp. 1165–1166.

50. Ibid., pp. 1169–1170.

51. For the decline of focal infection theory see Hughes, "Focal Infection Revisited," p. 375.

6. The Disappearance of Tic Illness

1. George B. Hassin, Arthur Stenn, and J. J. Burstein, "Stereotyped Acts or Attitude Tics? A Case with a Peculiar Anomaly of Gait," *The Journal of Nervous and Mental Disease,* 71 (1930): 27–32. See esp. pp. 28–30.

2. Hassin described Frank's tics as *"stereotyped,"* sometimes called "attitude tics." Stereotyped tics were more "complex" and manifested a greater "variety" than nonstereotyped tics, which, in contrast, were more impulsive and presented themselves at more regular intervals. Ibid., pp. 28–29. The term "stereotyped tics" had an old and often confusing history, going back at least to the French psychiatrist J. P. Falret, who in 1864 wrote that when an afflicted person "felt required to repeat" an act or a word "constantly, *exactly under the same form and with the same expressions,"* we can call this "completely stereotyped [stéréotypé]." J. P. Falret, *Des Maladies Mentales* (Paris: 1864), p. 193, italics in original.

3. Hassin, Stenn, and Burstein, "Stereotyped Acts or Attitude Tics? p. 30. A similar point was made in 1931 by two Italian physicians, who were

convinced that a patient's combined tic and Parkinson's were the result of multiple underlying causes including encephalitis and psychogenic factors. Marino Gopevich and Vittorio Romanin, "Su di una Varietà Rara di Sindroma Eccitomotoria Postencefilitca," *Il Policlinico*, 38 (1931): 763–766.

4. The strategy of isolating Frank for four months revealed the frustrations that many clinicians must have felt. Hassin seemed unsurprised that this approach "was of little avail and the patient insisted on leaving." Hassin, Stenn, and Burstein, "Stereotyped Acts or Attitude Tics?" p. 30.

5. New York Hospital/Cornell Archives. See folders for "convulsive tics," "habit spasms," and "chorea."

6. Chorea, box 21–29, 1926–1931, New York Hospital/Cornell Archives.

7. Case: 271065, 20 January to 12 February 1927, New York Hospital/Cornell Archives.

8. Case 295645, 21 October to 14 November 1930, ibid. There was also no evidence of rheumatic heart disease, a symptom generally associated with a diagnosis of chorea. Of course, rheumatic fever is not a necessary feature of Sydenham's chorea, but in the 1920s and 1930s most physicians believed this was so and looked for this as a sign of chorea. Among the other cases that obviously fit convulsive tics, but that were included as choreas are cases 300449, 18 August 1931, and 292304, April 1930.

9. Case 264866, 7 January to 2 February 1926, ibid.

10. Case 281230, 8 August to 7 September 1928, ibid.

11. Case 2927231, 28 January to 21 February 1931, ibid.

12. *Index-Catalogue of the Library of the Surgeon General's Office*, 2d ser., vol. 18 (Government Printing Office: Washington, D.C., 1913), pp. 261–265.

13. Ibid., vol. 10 (1932), p. 359.

14. Ibid., 4th ser., vol. 6 (1941), p. 291.

15. American Medical Association, *Quarterly Cumulative Index Medicus* 23 (1938): 1345; ibid., 24 (1938): 1274.

16. *Index-Catalogue of the Surgeon General's Office*, 5th ser., 3 (1961): 1555.

17. *Quarterly Cumulative Index Medicus* (1931), p. 302.

18. *Index-Catalogue of the Library of the Surgeon General's Office*, 4th ser., vol. 3 (Government Printing Office: Washington, D.C., 1938), pp. 600–609; see p. 606 for the listings of "mental manifestations."

19. Generoso Colucci, "Forme organiche e funzionali di spasmi e di tics," *La Riforma Medica*, 46 (7 July 1930): 1073–1078. Some saw tics as a combination of infectious and psychological factors. See G. Marañón, "Clinical Problems of Chorea, Tic, and Lack of Motor Coordination of a Growing Child: Differential Diagnosis," *Siglo Méd*, 97 (9 May 1936): 474–477.

20. Encephalitis was also connected with a number of other movement disorders. For instance, see Ladislas Benedik and Eugène de Thurzo, "Le

Rôle Inhibiteur des Excitations Périphériques sur les Tics Organiques," *Revue Neurologique*, 2 (1930): 701–703; F. von Bernuth, "Complicated Postencephalitic Tic in Children: Dilatation of the Heart in Postencephalitic State: 2 Cases," *Zeitschrift fur Kinder- und Jugendpsychiatrie*, 52 (1932): 534–555; C. I. Urechia and Mme. Retezeanu, "Parkinsonisme avec Torticolis Spasmodique ou avec Tics Buccaux," *Archives Internationales de Neurologie*, 56 (1937): 491–494; H. Claude, "Postencephalitic Parkinsonianism or Convulsive Tic With Coprolalia: 2 Cases," *Revue Général de Clinique et de Thérapie*, 53 (1939): 433–434; A. Huber and W. Österreicher, "Störungen der Zungenbewegung beim postenzephalitischen Parkinsonismus," *Med. Klinic*, 34 (1938): 470.

21. G. Heuyer [with comments by Drs. Vogt, Lautman, and Stern], "Présentation de Deux Jumelles Tics Encéphalitiques chez l'Une," *Annales Médico-Psychologiques*, 94 (1936): 228–223; quotation from p. 223.

22. Ibid., p. 234.

23. Lucien Cornil, "Les Rapports des Tics Convulsifs et de l'Hérédo-Syphilis," *Marseille-Médical*, 76 (1939): 818–822; quotation from pp. 819–820, italics in original.

24. Ibid., pp. 821–822.

25. Ibid., p. 821.

26. Mildred Creak, E. Guttman, and M. D. Munich, "Chorea, Tics, and Compulsive Utterances," *Journal of Mental Science*, 81 (1935): 834–839; quotation from p. 834.

27. H. Steck, "Les Syndromes Mentaux Postencéphalitiques," *Schweizer Archiv für Neurologie und Psychiatrie*, 27 (1931): 137–139, quoted in Creak, "Chorea, Tics, and Compulsive Utterances," p. 834.

28. B. Schultz, "Beiträge zur Genealogie der Chorea minor," *Zeitschr. fur d. ges. Neurol. und Psychiat.*, 117 (1928): 288–297.

29. Creak, Guttman, and Munich, "Chorea, Tics, and Compulsive Utterances," p. 835.

30. P. Guiraud, "Analyse du Symptome Stéréotypie," *L'Encephale*, 31 (1936): 229–269. Guiraud concluded that the term "stéréotypé" had become so confused that he warned that his "long study should convince a reader that one should no longer continue to use the term *stereotypes*" without the precision needed to differentiate what appeared to be a "collection of different [types of] symptoms" (p. 268).

31. Guiraud, "Analyse du Symptome Stéréotypie," pp. 251–253, 262–263.

32. Ibid., pp. 236–246.

33. For instance see B. Halporn, "Psychoneurosis Resembling Chorea," *Pennsylvania Medical Journal*, 33 (1930): 683–684; Wolfgang Hochheimer, "Zur Psychologie des Choreatikers," *Journal für Psycholgie und Neurologie*, 47 (1936): 49–115; P. Migault, "Syndrome Choréique et Syndrome Maniaque," *Paris Médical*, 2 (27 September 1930): 272–279; A. Simonini, "Psychic and Personality Disturbances in Chorea Following Polyar-

ticular Rheumatism: A Clinical Study," *Clinical Pediatrics,* 18 (1936): 242–246.

34. For more on Williams, see Howard I. Kushner, *Self-Destruction in the Promised Land: A Psychocultural Biology of American Suicide* (New Brunswick, N.J., Rutgers University Press, 1989), pp. 556–558.

35. Tom A. Williams, "Abnormal Movements (Tic): Nature and Treatment," *International Journal of Medicine and Surgery,* 45 (1932): 101–103; quotation from p. 103.

36. William Emet Blatz and Mabel Crews Ringland, *A Study of Tics in Pre-School Children* (Toronto: University of Toronto Press, 1935).

37. They cited a study published earlier (1928) by Blatz and another colleague, Helen Bott. See William Emet Blatz and Helen Bott, *Parents and the Preschool Child* (Toronto: J. M. Dent and Sons, 1928), p. 86, quoted in Blatz and Ringland, *Study of Tics in Pre-School Children,* pp. 15–16.

38. Ibid., pp. 10, 13.

39. Mary Chadwick, *Difficulties in Childhood Development* (New York: John Day Co., 1928), pp. 166–172; quoted in Blatz and Ringland, *Study of Tics in Pre-School Children,* p. 16. Chadwick's textbook was revised and republished in 1937, but Chadwick's views of tics remained unaltered. See Mary Chadwick, *Difficulties in Child Development,* new and abridged ed. (London: G. Allen & Unwin, 1937).

40. Annie Dolman Inskeep, *Child Adjustment in Relation to Growth and Development* (New York: D. Appleton, 1930), pp. 252–254. Also cited and discussed in Blatz and Ringland, *Study of Tics in Pre-School Children,* pp. 19–20.

41. Ibid., pp. 49–50, 54–55.

42. Each case history listed both the environmental and constitutional factors that may have contributed to the child's movements. Boenheim reported that 100 percent of those with tic had some form of psychiatric disorder, compared to only 36 percent of those with chorea. On the other hand, 55 percent of choreic children had rheumatic fever and/or heart disease, while none of the ticcers presented with these organic disorders. Curt Boenheim, "Über den Tic im Kindesalter," *Klinische Wochenschrift,* 9 (1930): 2005–2011.

43. Ibid., p. 2010. Once this was understood, Boenheim believed, clinicians would realize that they should turn to "depth psychology" as the best way to combat these compulsive symptoms. Similarly, the Austrian (Linz) neuropsychiatrist Georg Stiefler was persuaded that "it would be wrong to . . . explain all tics in the wide and narrow sense as organic." Rather, "there were three forms of tics: mental tic, hysterical forms of tics, and the striatal tics." Georg Stiefler, "Ueber die Tics," *Wiener Klinische Wochenschrift,* 48 (1935): 1439–1443; quotation from p. 1443. Other psychoanalysts focused more directly on some of the most florid (and sexually explicit) compulsive symptoms. For instance, in 1936 a Viennese

psychoanalyst elaborated a theory of coprolalia that subsumed it entirely in oedipal conflicts. Edmund Bergler, "Obscene Words," *Psychoanal. Q.*, 5 (1936): 226–248. Other representative articles taking the psychoanalytic approach in the 1930s include Hans Krisch, "Zur Theorie der Impuls- und Zwangshandlungen: Dargestellt an der Hand eines Falles von Tic und generalisierten, z.T. zwangshaft ausgebauten Zertrümmerungsimpulsen," *Zeitschrift für die gesamte Neurologie und Psychiatrie* 130 (1930): 257–295; and Yrjö Kulovesi, "Tic'in kehityksestä" [Tic Psychogenesis], *Duodecim*, 52 (1936): 137–154; E. Laszlo, "Psychotherapy for Tic," *Vie Médecine*, 19 (1938): 565–567.

44. Oliver Spurgeon English and Gerald H. J. Pearson, *Common Neuroses of Children and Adults* (New York: W. W. Norton, 1937), pp. 130–138.

45. Ibid., p. 137.

46. Ibid., pp. 134–135.

47. Instead, "a reconstruction of the psychosexual life is required" and "the most advisable method of treatment is psychoanalysis." But because "the intrapsychic dynamics of a tic involve very deep levels of the unconscious, . . . these cases require a very long and intensive psychoanalysis." Ibid., p. 138.

48. Andre DeWulf and Ludo van Bogaert, "Études Anatomo-Cliniques de Syndrômes Hypercinétiques Complexes," *Monatsschrift für Psychiatrie und Neurologie*, 104 (1941): 53–61; quotation from p. 59.

49. Ibid., p. 54. For the original see Georges Gilles de la Tourette, "La Maladie des Tics Convulsifs," *La Semaine Médicale*, 19 (1899): 153–156.

50. George Gilles de la Tourette, "Étude sur une Affection Nerveuse Caractérisée par de l'Incoordination Motrice Accompangnée d'Écholalie et de Coprolalie (Jumping, Latah, Myriachit)," *Archives de Neurologie*, 9 (1885): 19–42, 158–200, see p. 200.

51. DeWulf and van Bogaert, "Études Anatomo-Cliniques de Syndrômes Hypercinétiques Complexes," pp. 53–54.

52. Ibid., pp. 55–58.

53. A central nervous system depressant that restricts the transmission of choline.

54. Ibid., p. 58.

55. Ibid., pp. 59–60.

56. "Nothing" had been uncovered since then "that had authorized reconsideration of this illness as an infection which becomes integrated with illnesses of the extra-pyramidal system or as a post-infectious sequel." Ludo van Bogaert, "Maladie des Tics de Gilles de la Tourette," *Traité de Médecine*, 16 (1949): 246–248; quotation from p. 248.

57. For more on this see Ellen Herman, *The Romance of American Psychology: Political Culture in the Age of Experts* (Berkeley: University of California Press, 1995), pp. 65–81.

7. Margaret Mahler and the Tic Syndrome

1. Sandor Ferenczi, "Psycho-Analytical Observation on Tic," *International Journal of Psycho-Analysis,* 2 (1921): 1–30; quotation from pp. 2–4 (italics in original).

2. Curt Boenheim, "Über den Tic im Kindesalter," *Klinische Wochenschrift,* 9 (1930): 2005–2011.

3. Oliver Spurgeon English and Gerald H. J. Pearson, *Common Neuroses of Children and Adults* (New York: W. W. Norton, 1937), pp. 130–138.

4. Joseph Wilder, "Ein Fall von Maladie des Tics," in S. Karger, ed. *Beitrage zum Ticproblem* (Berlin: 1927), pp. 96–107.

5. See for instance Josef Wilder, "Der Tic Convulsif," *Jahresklirse für Ärztliche Fortbildung,* 21 (1930): 13–27.

6. Wilder, who had written extensively on convulsive tics in the late 1920s and early 1930s, had fled Germany for the United States in the late 1930s. By the mid-1940s he was writing articles in English with almost the exact words that he had used in his earlier German articles about convulsive tics. See Joseph Wilder, "Tic Convulsif as a Psychosomatic Problem," *The Nervous Child,* 5 (1946): 365–371; quotation from p. 371.

7. Ibid., pp. 365–366.

8. Ibid., p. 370.

9. Ibid., p. 371, italics in original.

10. Margaret S. Mahler, *The Memoirs of Margaret S. Mahler,* compiled and edited by Paul E. Stepansky (New York: The Free Press, 1988), esp. pp. 11–18, 58–72; Louise J. Kaplan, *Oneness and Separateness: From Infant to Individual* (New York: Simon and Schuster, 1978), pp. 15–19.

11. Vilma Kovács, "Analyse eines Falles von 'Tic Convulsif,'" *Internationale Zeitschrift für Psychoanalyse,* 11 (1925): 318–325.

12. Margaret S. Mahler and Leo Rangell, "A Psychosomatic Study of Maladie des Tics (Gilles de la Tourette's Syndrome)," *The Psychiatric Quarterly,* 17 (1943): 579–603. Quotation on p. 579. Mahler and Rangell used "maladie des tics," "convulsive tics," and "Gilles de la Tourette's disease" as synonyms.

13. Ibid., p. 593. Mahler not only had read Gilles de la Tourette, but also was familiar with Charcot's, Guinon's, and Meige and Feindel's views.

14. Ibid., p. 579; I. W. Wechsler, *A Textbook of Clinical Neurology,* 3rd ed. (Philadelphia: W. B. Saunders Co., 1937), p. 619, quoted in Mahler and Rangell, "A Psychosomatic Study of Maladie des Tics," p. 580.

15. Margaret S. Mahler, Jean A. Luke, and Wilburta Daltroff, "Clinical and Follow Up Study of the Tic Syndrome in Children," *American Journal of Orthopsychiatry,* 15 (1945): 631–647.

16. Ibid., p. 633.

17. Ibid. (italics added).

18. Mahler and Rangell, "A Psychosomatic Study of Maladie des Tics," p. 580.

19. Kaplan, *Oneness and Separateness*, p. 15.

20. Mahler and Rangell, "A Psychosomatic Study of Maladie des Tics," p. 582.

21. Ibid., pp. 592–593.

22. Ibid., pp. 581, 583–584, 593. Mahler and Rangell decided to omit fully reporting the psychological analysis of Freddie's inkblot test. The conclusions were published two years later by Margaret Naumburg, an art therapist hired to work with Freddie. Margaret Naumburg, "The Psychodynamics of the Art Expression of a Boy Patient with Tic Syndrome," *The Nervous Child*, 4 (1945): 374–409, see pp. 378–379. Naumburg's essay, along with one including Freddie's Rorschach, appeared later in a volume edited by Mahler.

23. Mahler and Rangell, "A Psychosomatic Study of Maladie des Tics," pp. 581, 588.

24. Ibid., pp. 585, 587.

25. Ibid., pp. 586–588.

26. Ibid., p. 600. Margaret Naumburg, Freddie's art therapist, published an extensive article on Freddie's therapy, including thirty samples of his drawings, in Mahler's edited 1945 volume. Naumburg, "The Psychodynamics of the Art Expression," pp. 407–408.

27. Margaret Schoenberger Mahler, "Tics and Impulsions in Children: A Study of Motility," *The Psychoanalytic Quarterly*, 13 (1944): 430–444.

28. Ibid., pp. 435–436.

29. *Tics in Children*, ed. with an introduction by Margaret Schoenberger Mahler, *The Nervous Child*, 4 (1945): 306–419.

30. Margaret Schoenberger Mahler and Irma L. Gross, "Psychotherapeutic Study of a Typical Case with Tic Syndrome," *The Nervous Child*, 4 (1945): 359–373.

31. The paper was to have been presented to an earlier canceled psychiatric meeting. Mahler, Luke, and Daltroff, "Clinical and Follow Up Study of the Tic Syndrome in Children," pp. 631–647.

32. Margaret Schoenberger Mahler, "Introductory Remarks," *The Nervous Child*, 4 (1945): 307.

33. Samuel Ritvo, "Survey of the Recent Literature on Tics in Children," *The Nervous Child*, 4 (1945): 308–312. Ritvo, who would become one of Mahler's most devoted followers, had shared an office with her at the Psychiatric Institute.

34. Mahler, Luke, and Daltroff, "Clinical and Follow Up Study of the Tic

Syndrome in Children," pp. 631–647, quoted by Ritvo, "Recent Literature on Tics in Children," pp. 310–311.

35. Ibid., p. 311.

36. Mahler and Rangell, "A Psychosomatic Study of Maladie des Tics,"
    p. 584.

37. Bernard L. Pacella, "Physiologic and Differential Diagnostic Considera-
    tions of Tic Manifestations in Children," *The Nervous Child*, 4 (1945):
    313–317.

38. Thesi Bergmann, "Observation of Children's Reactions to Motor Re-
    straint," *The Nervous Child*, 4 (1945): 318–328. Bergmann, a loyal fol-
    lower, endorsed Mahler's assertions, but also raised the interesting ques-
    tion of why her own [Bergmann's] patients "who displayed such severe
    temper tantrums, did not go all the way through to the development of
    tics" (p. 328).

39. Paul H. Hoch, "Neurodynamics of Tics," *The Nervous Child*, 4 (1945):
    329–334.

40. Hoch believed that some tics were purely psychogenic. Among them was
    what Hoch labeled "maladie des tics," a category Hoch neither defined
    nor attached to the earlier literature, but presented as if the classification
    were widely accepted. Ibid., pp. 333–334.

41. Esther Menaker, "Hypermotility and Transitory Tic in a Child of Seven,"
    The Nervous Child, 4 (1945): 335–341; quotations from pp. 335–337.
    Menaker and her husband had been sponsors for Mahler and other psy-
    choanalysts fleeing Hitler's Europe. The Mahlers and Menakers became
    close social friends. Mahler, *Memoirs*, pp. 117–118.

42. Menaker, "Hypermotility and Transitory Tic in a Child of Seven,"
    pp. 338–340. Support for this was provided by Alice's teacher who re-
    ported "a great deal of sexual play in the school situation. Several little
    girls, of whom Alice was one, were unable to sit still in the classroom.
    They would run out and congregate in the bathroom, where they indulged
    in 'secret games'" (p. 339).

43. Ibid., p. 340. Menaker suggested, despite any other substantive evidence,
    that Alice "had experienced some form of sexual seduction—perhaps at
    the hands of her brother—as a much younger child."

44. Ibid., p. 341.

45. Zygmunt A. Piotrowski, "Rorschach Records of Children with a Tic Syn-
    drome," *The Nervous Child*, 4 (1945): 342–352.

46. Ruth Henning Latimer, "The Parent-Child Relationships in Children
    Afflicted With Tics," *The Nervous Child*, 4 (1945): 353–358; quotation
    from p. 355.

47. It also contained art therapist Margaret Naumburg's essay on Freddie.
    Naumburg, "The Psychodynamics of the Art Expression," pp. 374–409.

48. Mahler and Gross, "Psychotherapeutic Study of a Typical Case," p. 359.
49. Ibid., p. 363. Laboratory and neurological examinations revealed no organic problems. Pete was sedated during his seven-and-a-half week hospital stay, but the sedatives had no effect on his tics. His doctors also attempted to interview Pete under hypnosis induced by sodium amytal. This procedure only caused Pete to fall asleep.
50. Ibid., pp. 361–362.
51. Ibid., pp. 363–364.
52. Ibid., p. 365.
53. Ibid., p. 367. When Pete complained, "As long as you don't give me any 'real' treatment with medicine, it [the tics] will always get better in the hospital and worse again as soon as I go home." Mahler and Gross believed that they had hit psychological pay dirt: "Using his own statement that the tics were bound to increase at home, it was pointed out to him that something in the home environment must have to do with the tics' exacerbation" (p. 366).
54. Ibid., p. 369.
55. Ibid., pp. 371–372. Pete was also the main subject of Piotrowski's essay on Rorschach, which concluded that Pete's "Rorschach record reveals many tensions and conflicting tendencies which might be responsible for the tics." Piotrowski, "Rorschach Records of Children with a Tic Syndrome," pp. 350–351.
56. Mahler, Luke, and Daltroff, "Clinical and Follow Up Study of the Tic Syndrome in Children," pp. 631–647.
57. Ibid., pp. 634–635.
58. Ibid., pp. 646–647.
59. Ibid., p. 646.
60. Margaret S. Mahler and Jean A. Luke, "Outcome of the Tic Syndrome," *Journal of Nervous and Mental Diseases,* 103 (1946): 433–445.
61. Mahler and Luke, "Outcome of the Tic Syndrome," pp. 435, 440–441.
62. Ibid, pp. 442–443.
63. Ibid., p. 445.
64. Margaret Schoenberger Mahler, "A Psychoanalytic Evaluation of Tic in the Psychopathology of Children: Symptomatic and Tic Syndrome," *Psychoanalytic Study of the Child* (1949): 279–310.
65. Ibid., p. 307.
66. Mahler's taped reflections were edited and compiled by Paul E. Stepansky and published in 1988 as *The Memoirs of Margaret S. Mahler.*
67. Mahler, *Memoirs,* p. 111.
68. Arthur K. Shapiro, Elaine S. Shapiro, Ruth D. Bruun, and Richard D. Sweet, *Gilles de la Tourette Syndrome* (New York: Raven Press, 1978), p. 63.
69. Mahler, *Memoirs,* p. 4.

8. Haloperidol and the Persistence of the Psychogenic Frame

1. J. E. Heuscher, "Intermediate State of Consciousness in Patients with Generalized Tics," *Journal of Nervous and Mental Disease,* 117 (1953): 29–38; quoted in Richard P. Michael, "Treatment of a Case of Compulsive Swearing," *British Medical Journal,* 1 (1957): 1506–1508; quotation from p. 1507. Heuscher's original wording actually was that the tic was the "last desperate attempt of the patients to find a compromise solution between their inner conflicts and the exigencies of their environment without becoming psychotic." Also citing and endorsing this view were Leon Eisenberg, Eduard Ascher, and Leo Kanner, "A Clinical Study of Gilles de la Tourette's Disease (Maladie des Tics) in Children," *Journal of American Psychiatry,* 115 (1959): 715–723, see p. 722.

2. Margaret W. Gerard drew heavily from Mahler's publications, endorsing the view that tics served as "a partial substitution for impulse expression," while simultaneously maintaining "parental love by a nontaboo action." Margaret W. Gerard, "The Psychogenic Tic in Ego Development," *Psychoanalytic Study of the Child,* 2 (1946): 141–161.

3. Eduard Ascher, "Psychodynamic Considerations in Gilles de la Tourette's Disease (Maladie des Tics)," *American Journal of Psychiatry,* 105 (1948): 267–276; quotation from p. 275. For a similar view see Avery D. Weisman, "Nature and Treatment of Tics in Adults," *A.M.A. Archives of Neurology and Psychiatry,* 68 (1952): 444–459.

4. Z. Alexander Aarons, "Notes on a Case of *Maladie des Tics,*" *Psychoanalytic Quarterly,* 27 (1958): 194–204; quotations from pp. 199, 202–204.

5. Other examples of this approach are found in William G. Tobin and John B. Reinhart, "Tic de Gilles de la Tourette," *American Journal of Diseases of Children,* 101 (1961): 778–783; Robert W. Downing, Nathan L. Comer, and John N. Ebert, "Family Dynamics in a Case Study of Gilles de la Tourette's Syndrome," *Journal of Nervous and Mental Disease,* 138 (1964): 548–557.

6. Behavior modification was the other psychological intervention attempted in the 1950s and 1960s. Although it was often used as an adjunct to other therapies, sometimes behavior modification was the only intervention attempted. For reports claiming success via this method, see D. Walton, "Experimental Psychology and the Treatment of a Ticqueur," *Journal of Child Psychology and Psychiatry,* 2 (1961): 148–155; A. Abi Rafi, "Learning Theory and the Treatment of Tics," *Journal of Psychosomatic Research,* 6 (1962): 71–76; H. Milton Erickson, "Experimental Hypnotherapy in Tourette's Disease," *The American Journal of Clinical Hypnosis,* 7 (1965): 325–331; D. F. Clark, "Behaviour Therapy of Gilles de la Tourette's Syndrome," *British Journal of Psychiatry,* 112 (1966): 771–

778; J. P. Sichel and R. Durand de Bousingen, "Le Traitement des Tics chez l'Enfant par le Training Autogene," *Revue de Neuropsychiatrie Infantile*, 15 (1967): 931–937. A Utah psychiatrist claimed to have success in managing the symptoms of Tourette syndrome in a fourteen-year-old boy through group therapy. Eugene J. Faux, "Gilles de la Tourette Syndrome," *Archives of General Psychiatry*, 14 (1966): 139–142.

7. P. F. Girard and Schott, "Le Mécanisme Physio-Pathologique des Tics de la Gaucherie au Tic," *Pédiatrie* (1948): 2–14; see p. 20.

8. Claude Launay, "Les Tics chez l'Enfant," *Revue du Praticien*, 4 (1954): 2783–2790, see esp. p. 2790.

9. M. R. Lanter, "Traitement des Tics chez l'Adulte par le Parpanit, Associe à la Narcoanalyse," *Annale Médico-Psychologique*, 2 (1950): 277–278. The patient was given 150 to 200 milligrams per day of parpanit for three months. Lanter used Janet's method of "narco-analyse," in which suggestion was purportedly achieved through use of chemical agents, rather than through hypnosis.

10. Raymond de Saussure, "Discussion sur l'Étiologie d'un Tic survenu Quinze Mois après une Encéphalite Léthargique Atypique," *Schweizer Archiv für Neurologie und Psychiatrie*, 12 (1923): 294–317; Erwin Straus, "Über die organische Natur der Tics und der Koprolalie," *Zentralblatt für die gesamte Neurologie und Psychiatrie*, 47 (1927): 698–699. Also see his "Untersuchungen Über die postchoreatischen Motilitätsstörungen, insbesondere die Beziehungen der Chorea minor zum Tic," *Monatsschrift für Psychiatrie und Neurologie*, 66 (1927): 261–324.

11. G. de Morsier, "Les Tics et la Maladie des Tics Convulsifs (Gilles de la Tourette)," *Médicine et Hygiene*, 9 (1951): 260–261.

12. Ibid., pp. 260–261.

13. Ibid., p. 261.

14. François Thiébaut, "Introduction à l'Étude des Mouvement Involuntaires," *Revue Neurologique*, 86 (1952): 535–548, see esp. pp. 543–548. The caudate nucleus, putamen, and globus pallidus are sometimes referred to as the corpus striatum (or striate cortex). Historically, including in most discussions of Tourette syndrome, the basal ganglia and the corpus striatum have been used interchangeably. For more see Francis Schiller, "The Vicissitudes of the Basal Ganglia," *Bulletin of the History of Medicine*, 41 (1967), 515–538.

15. M. B. Schmitt, "Effets d'un Curarisant de Synthèse sur les Tics Psychomoteurs," *Annales Médico-Psychologique*, 1 (1953): 102.

16. Franco Giberti, "Tics Multipli e Malattia di Gilles de la Tourette," *Sistema Nervoso*, 2 (1954): 136–153; quotation from p. 151. Also see W. P. Mazur, "Gilles de la Tourette's Syndrome," *Canadian Medical Association Journal*, 69 (1953): 520–522.

17. Johan Ludwig Clauss and Karl Balthasar, "Zur kenntnis der generalisier-

ten Tic Krankheit (Maladie des Tics, Gilles de la Tourette'sche Krankheit)," *Archiv für Psychiatrie und Nervenkrankheiten,* 191 (1954): 389–418.

18. Karl Balthasar, "Über das anatomische Substrat der generalisierten Tic-Krankheit (Maladie des Tics, Gilles de la Tourette): Entwicklungshemmung des Corpus Striatum," *Archiv für Psychiatrie und Nervenkrankheiten,* 195 (1957): 531–549; see esp. pp. 540–541, 548.

19. See Curt Boenheim, "Über den Tic im Kindesalter," *Klinische Wochenschrift,* 9 (1930): 2005–2011; Oliver Spurgeon English and Gerald H. J. Pearson, *Common Neuroses of Children and Adults* (New York: W. W. Norton, 1937), pp. 130–138; Joseph Wilder, "Tic Convulsif as a Psychosomatic Problem," *The Nervous Child,* 5 (1946): 365–371, esp. p. 370; Margaret S. Mahler, Jean A. Luke, and Wilburta Daltroff, "Clinical and Follow Up Study of the Tic Syndrome in Children," *American Journal of Orthopsychiatry,* 15 (1945): 631–647, see p. 633.

20. See J. Krasowska and E. Cyran, "O Zespole Gilles de la Tourette'a," *Neurologia, Neurochirurgia I Psychiatria Polska,* 16 (1966): 1367–1372; M. Subramaniam, "The Gilles de la Tourette Syndrome," *Medical Journal of Malaya,* 21 (1966): 95–96. An Australian physician, R. M. Ellison, was convinced that a sixty-seven-year-old man's Tourette syndrome was triggered by a severe episode of pneumonia when he was five years old. R. M. Ellison, "Gilles de la Tourette's Syndrome," *Medical Journal of Australia,* 51 (1964): 153–155.

21. Laurence Selling, "The Role of Infection in the Etiology of Tics," *Archives of Neurology and Psychiatry,* 22 (1929): 1163–1171.

22. Edward E. Brown, "Tics (Habit Spasms) Secondary to Sinusitis," *Archives of Pediatrics,* 74 (1957): 39–46.

23. Ibid., pp. 39, 41–43.

24. Ibid., pp. 39, 43–44.

25. Ibid., pp. 44–45.

26. Ibid., p. 44.

27. L. J. Meduna, "The Mode of Action of Carbon Dioxide Treatment in Human Neuroses," reprinted in Edward Podolsky, ed., *The Neuroses and Their Treatment* (New York: Philosophical Library, 1958), pp. 525–528.

28. Meduna had been one of the prime advocates of treating depressive patients by inducing convulsions. He advocated the use of a camphor-like drug, metrazol, rather than insulin, which prior to electroshock, had been the preferred mechanism to induce "therapeutic" convulsions. For more on Meduna and convulsive therapies see Elliot S. Valenstein, *Great and Desperate Cures: The Rise and Decline of Psychosurgery and Other Radical Treatments for Mental Illness* (New York: Basic Books, Inc., 1986), pp. 45–48.

29. L. J. Meduna, "The Carbon Dioxide Treatment: A Review," reprinted in

Podolsky, ed., *The Neuroses and Their Treatment*, pp. 502–524; see esp. pp. 504, 521–522.

30. Ibid., p. 524.

31. Quoted in ibid., p. 513.

32. Michael, "Treatment of a Case of Compulsive Swearing," p. 1507.

33. Preston W. DeShan, "Gilles de la Tourette's Syndrome: Report of a Case," *Journal of the Oklahoma State Medical Association*, 54 (1961): 636–638; George Challas, James L. Chapel, and Richard L. Jenkins, "Tourette's Disease: Control of Symptoms and Its Clinical Course," *International Journal of Neuropsychiatry*, 3 (1967): 95–109, see p. 102. A 1969 British study asserted that the cause of Tourette syndrome was psychological, but conceded that a review of 180 cases revealed that psychotherapy was no more effective than CO_2 or doing nothing. J. A. Corbett, A. M. Mathews, P. H. Connell, and D. A. Shapiro, "Tics and Gilles de la Tourette's Syndrome: A Follow-Up Study and Critical Review," *British Journal of Psychiatry*, 115 (1969): 1229–1241, esp. p. 1240.

34. Ian J. MacDonald, "A Case of Gilles de la Tourette Syndrome, with Some Aetiological Observations," *British Journal of Psychiatry*, 109 (1963): 206–210; quotations from pp. 207–208.

35. Ibid., p. 208. There were also reports of failures with CO_2 inhalations. See Demtri J. Polites, David Kruger, and Ian Stevenson, "Sequential Treatments in a Case of Gilles de la Tourette's Syndrome," *British Journal of Medical Psychology*, 38 (1965): 43–52, see p. 48.

36. For background see Jack Pressman, *Last Resort: Psychosurgery and the Limits of Medicine* (Cambridge: Cambridge University Press, 1998), esp. pp. 194–235; Valenstein, *Great and Desperate Cures*, pp. 220–252.

37. Pressman, *Last Resort: Psychosurgery and the Limits of Medicine*, pp. 337–339. Watts eventually ceased participating in transorbital lobotomies, while Freeman devoted his career to performing them.

38. Reported in Harold Stevens, "The Syndrome Gilles de la Tourette and its Treatment," *Medical Annals of the District of Columbia*, 33 (1964): 277–279; quotations from p. 279.

39. E. F. W. Baker, "Gilles de la Tourette Syndrome Treated by Bimedial Frontal Leucotomy," *Canadian Medical Association Journal*, 86 (1962): 746–747; quotations from p. 747. For the monkey study, see E. L. Foltz, L. M. Knopp, and A. A. Ward, Jr., "Experimental Spasmodic Torticollis," *Journal of Neurosurgery*, 16 (1959): 55–72.

40. Uta Asam and W. Karrass, "Gilles de la Tourette-Syndrom und Psychochirurgie," *Acta Paedopsychiatrica*, 47 (1981): 39–48; quotations from p. 47.

41. Solomon H. Snyder, *Drugs and the Brain* (New York: Scientific American Library, 1996), pp. 70–72; Nancy C. Andreasen, *The Broken Brain: The Biological Revolution in Psychiatry* (New York: Harper & Row, 1985), pp. 192–193.

42. Michael, "Treatment of a Case of Compulsive Swearing," p. 1507.
43. MacDonald, "A Case of Gilles de la Tourette Syndrome, with Some Aetiological Observations," p. 210.
44. Stevens, "The Syndrome Gilles de la Tourette and its Treatment," p. 279.
45. Baker, "Gilles de la Tourette Syndrome Treated by Bimedial Frontal Leucotomy," p. 747.
46. S. Bockner, "Gilles de la Tourette's Disease," *Journal of Mental Science,* 105 (1959): 1078–1081; quotation from p. 1080.
47. Ibid., pp. 1080–1081.
48. Alvin M. Mesnikoff, "Three Cases of Gilles de la Tourette's Syndrome Treated with Psychotherapy and Chlorpromazine," *A.M.A. Archives of Neurology and Psychiatry,* 81 (1959): 710.
49. Eisenberg, Ascher, and Kanner, "A Clinical Study of Gilles de la Tourette's Disease," p. 722.
50. Jack Rapoport, "Maladie des Tics in Children," *American Journal of Psychiatry,* 115 (1959): 177–178; Leon Eisenberg, Eduard Ascher, and Leo Kanner, "Reply to Dr. Rapoport," ibid., p. 179; Doris H. Milman, "Multiple Tics," ibid., 116 (1960): 935–936; Jerome M. Schneck, "Gilles de la Tourette's Disease," ibid., 117 (1960): 78. Also see James R. Dunlap, "A Case of Gilles de la Tourette's Disease (Maladie des Tics): A Study of the Intrafamily Dynamics," *Journal of Nervous and Mental Disease,* 130 (1960): 340–344.
51. P. J. F. Walsh, "Compulsive Shouting and Gilles de la Tourette's Disease," *British Journal of Clinical Practice,* 16 (1962): 651–655.
52. Roman Dolmierski and Maria Kloss, "De la Maladie de la Tourette," *Annales Médico-Psychologiques,* 129 (1962): 225–232.
53. W. Sulestrowski and L. Wdowiak, "Des Syndromes Névrosiques Organiquement Conditionnés," *Polski Tygodnik Lekarski,* 40 (1957), quoted in Dolmierski and Kloss, "De la Maladie de la Tourette," pp. 227–228.
54. She also began to have Parkinson's symptoms. Ibid., p. 231.
55. Ibid. Other antipsychotic agents, including the antianxiety drug trifluoperazine [Stelazine], were claimed to be effective in controlling tics and vocalizations. See Polites, Kruger, and Ian, "Sequential Treatments in a Case of Gilles de la Tourette's Syndrome," pp. 50–51.
56. See Snyder, *Drugs and the Brain,* pp. 77, 81–83; 86–87.
57. In Europe haloperidol's trade name is Serenase.
58. Jean N. Seignot, "Un Cas de Maladie des Tics de Gilles de la Tourette Guéri par le R.1625," *Annales Médico- Psychologiques,* 119 (1961): 578–579.
59. Seignot updated the patient's reaction to haloperidol in a paper prepared for an American Psychiatric Association meeting in 1970. "A few years later," Seignot reported, "the tics became more frequent, in spite of the continued treatment of 0.6 mg per day. The dose of haloperidol was finally

increased to 3 mg per day, and this made the tics disappear completely. This treatment has been continued regularly at this dose for several years, and is perfectly tolerated with the sole addition of chlorhydrate of heptaminol to counteract the feeling of lassitude by the quantity of haloperidol." Jean N. Seignot, "A Case of Gilles de la Tourette's Disease After 10 Years' Treatment with Haloperidol (R.1625)," in Faruk S. Abuzzahab and Floyd O. Anderson, ed., *Gilles de la Tourette's Syndrome* (St. Paul: Mason Publishing, 1976), pp. 159–162; quotation from p. 161.

60. Seignot, "Un Cas de Maladie de Tics de Gilles de la Tourette Guéri par le R.1625," pp. 578–579. Almost simultaneously, an Italian team reported similar results. While this study would often be cited in studies that appeared after 1968, it was not as well-known or influential as Seignot's report. See G. Caprini and V. Melotti, "Una Grave Sindrome Ticcosa Guarita con Halopéridol," *Revista Sperimentale Freniatra*, 85 (1961): 191–196.

61. George Challas and William Brauer, "Tourette's Disease: Relief of Symptoms with R.1625," *American Journal of Psychiatry*, 120 (1963): 283–284.

62. Challas and Brauer's follow-up appeared in their colleagues James L. Chapel, Noel Brown, and Richard L. Jenkins's article, "Tourette's Disease: Symptomatic Relief with Haloperidol," *American Journal of Psychiatry*, 121 (1964): 608–610; quotation from p. 609.

63. Ibid., p. 609.

64. Ibid., p. 610.

65. In fact, they could find no suitable American journal willing to publish their article. Interview with Dr. Elaine S. Shapiro, Scarsdale, New York, 13 June 1996. Tape and transcript in the author's possession.

66. Alexander R. Lucas, "Gilles de la Tourette's Disease in Children: Treatment with Phenothiazine Drugs," *American Journal of Psychiatry*, 121 (1964): 607–608. Soon, convinced by the superior and seemingly more focused action of haloperidol, Lucas experimented with it on two new patients, reporting that symptom reduction in both "has been excellent." Alexander R. Lucas, "Gilles de la Tourette's Disease in Children: Treatment with Haloperidol," *American Journal of Psychiatry*, 124 (1967): 243–245.

67. U. Perini and B. Lampo, "L'Haloperidol nella Terapia delle Sindromi Ticcose," *Minerva Medica*, 55 (1964): 3387–3390; quotation from p. 3389.

68. T. Sobierski, "A Propos de la Maladie de Gilles de la Tourette," *Sem. Thérapeutic*, 40 (1964): 129. Also see Nuala M. Healy and Ingo Fischer, "Gilles de la Tourette Syndrome in an Autistic Child," *Journal of the Irish Medical Association*, 57 (1965): 93–94.

69. Diane H. Kelman, "Gilles de la Tourette's Disease in Children: A Review of the Literature," *Journal of Child Psychology and Psychiatry*, 6 (1965):

219–226. Kelman examined a number of issues including sex, age of onset, form of the development of symptoms, family and personal history
of the patients, IQ, and neurological and EEG examinations. Kelman's
epidemiological findings were consistent with her predecessors in that she
found that the symptoms appeared more frequently in males than in females; the age of onset was ten years or younger; the sequence of symptoms generally followed that outlined by Gilles de la Tourette, although
"the prognosis for this syndrome is not as poor as de la Tourette himself
predicted" (p. 255). So far as she could determine from the literature,
intelligence was normally distributed, but there was insufficient information to draw any conclusions about familial patterns. Although she included some psychoanalytic studies, Kelman specifically excluded Margaret Mahler's study on the grounds that Mahler and her colleagues had
no clear or consistent method for distinguishing patients with Tourette's
from "tiqueurs in general" (p. 220n).

70. See table, "Treatment and Outcome," ibid., p. 223.

71. Janice R. Stevens and Paul H. Blachly, "Successful Treatment of the
Maladie des Tics, Gilles de la Tourette's Syndrome," *American Journal
of Diseases in Children,* 112 (1966): 541–545.

72. Ibid., pp. 541–543,

73. Ibid., p. 543.

74. Ibid., pp. 543–545.

75. Ibid., pp. 543–544.

76. George Challas, James Chapel, and Richard L. Jenkins, "Tourette's Disease: Control of Symptoms and Its Clinical Course," *International Journal of Neuropsychiatry,* 3 (1967): 95–109, quotation from pp. 102–103.
One of the patients also received CO_2 inhalations.

77. James L. Chapel, "Gilles de la Tourette's Disease, the Past and the Present," *Canadian Psychiatric Association Journal,* 11 (1966): 324–329;
quotation from p. 327. Chapel added (pp. 327–328) that "it must be
emphasized that while the symptoms are very well controlled in all cases,
there is no reason to assume that the great hostility locked within these
patients is directly affected by the drug [Haloperidol] Psychotherapy,
counselling of parents and manipulation should be employed to tap and
drain this reservoir of anger."

78. Challas, Chapel, and Jenkins, "Tourette's Disease: Control of Symptoms
and Its Clinical Course," p. 103. In fact, they noted that "no failures with
haloperidol have been reported to date" (p. 97).

79. A. Kumento and M.-L. Koski, "Syndroma Gilles de la Tourette," *Acta
Paediatrica,* 177 (1967): 62–63. For a similar view see R. C. McKinnon,
"Gilles de la Tourette Syndrome: A Case Showing Electroencephalo-
Graphic Changes and Response to Haloperidol," *The Medical Journal of
Australia,* 54 (1967): 21–22. Others, citing their success with haloperidol,

concluded that Gilles de la Tourette's disease was most likely an organic disorder. See E. Boschi and G. C. La Maida, "Considerazioni Sulla Mallattia di Gilles de la Tourette," *Archivo Di Psicologia, Neurologia e Psichiatria,* 28 (1967): 141–160, see esp. pp. 147–151.

80. Kendall B. Corbin, J. Rodney Field, Norman P. Goldstein, and Donald W. Klass, "Further Observations on Tourette's Syndrome," *Proceedings of the Australian Association of Neurologists,* 5 (1968): 447–453; quotation from p. 447. Similarly, Essex (Britain) psychiatrist S. J. M. Fernando found that "on the basis of present evidence it appears that the management of a case must be broadly based. Therapy with phenothiazines or haloperidol should be accompanied by attempts to alleviate any social or psychological stress." S. J. M. Fernando, "Gilles de la Tourette's Syndrome," *British Journal of Psychiatry,* 113 (1967): 607–617; quotation from p. 614.

81. Hilde Bruch and Lawrence C. Thum, "Maladie des Tics and Maternal Psychosis," *Journal of Nervous and Mental Disease,* 146 (1968): 446–456.

82. Ibid., pp. 454–456.

83. Stanton Peele, "Reductionism in the Psychology of the Eighties: Can Biochemistry Eliminate Addiction, Mental Illness, and Pain?" *American Psychologist,* 36 (1981): 807–818.

84. This view is attributed to Thomas S. Kuhn, or at least the way social scientists have read him lately. See Thomas S. Kuhn, *The Structure of Scientific Revolutions* (Chicago: University of Chicago Press, 2d ed., 1970).

9. The French Resistance

1. Jean Delay, "Tic Organique et Tic Mental," *Annales Médico-Psychologiques,* 99 (1941): 328–333.

2. Julien Rouart, "Tics et Personnalité de l'Enfant," *L'Evolution Psychiatrique,* 1 (1947): 267–292.

3. Henri F. Ellenberger, *The Discovery of the Unconcious: The History and Evolution of Dynamic Psychiatry* (New York, Basic Books, 1970), pp. 346–347.

4. See Solomon H. Snyder, *Drugs and the Brain* (New York: Scientific American Library, 1996), pp. 71–73; Nancy C. Andreasen, *The Broken Brain: The Biological Revolution in Psychiatry* (New York: Harper & Row, 1985), pp. 192–193.

5. Delay, "Tic Organique et Tic Mental," p. 328.

6. "In a group of tics, . . . the largest number of them will be attached to material lesions of the nervous system . . . By appearing to advocate this

idea of tics, [I] will be put somewhat in contradiction with [my] earlier writings. New facts, that no one was able to anticipate, have forced this revision." Henry Meige quoted by Delay, in "Tic Organique et Tic Mental," p. 328. I have been unable to locate the original or to verify Delay's claim.

7. Ibid., pp. 328–329, italics in original.

8. Amblyopia is reduced vision in an eye that itself appears to be normal. The condition has a variety of predisposing causes including toxins, alcoholism, smoking, and demyelinating disorders.

9. Ibid., p. 330.

10. Pierre Janet, *Névroses et Idées Fixes,* 2 vols. (Paris: F. Alcan, 1898). See vol. 2, "par le dr. Fulgence Raymond [et] le dr. Pierre Janet," especially pp. 381–397. Charcot served as chairman of Janet's medical thesis committee. For more on Janet's career, see Ellenberger, *The Discovery of the Unconcious,* pp. 331–417, esp. pp. 340–344. Like Rouart, Delay trained under Janet and during the occupation the elderly Janet had treated a few patients each week at Delay's Paris clinic. Ibid., pp. 346–347.

11. Hypnotic suggestion was one of Janet's chief therapeutic interventions. For a useful discussion contrasting Freud's psychoanalysis with Janet's depth psychology, see Ian Hacking, *Rewriting the Soul: Multiple Personality and the Sciences of Memory* (Princeton: Princeton University Press, 1995), pp. 195–197.

12. Janet, *Névroses et Idées Fixes,* 2: 378.

13. For a discussion of Janet's view of obsessive-compulsive disorders, see Roger Pitman, "Pierre Janet on Obsessive-Compulsive Disorder (1903)," *Archives of General Psychiatry,* 44 (1987): 226–232.

14. Janet, quoted in Serge Lebovici, *Les Tics chez l'Enfant* (Paris: Presses Universitaires de France, 1951), p. 131.

15. Janet, *Névroses et Idées Fixes,* 1, quoted in Pitman, "Pierre Janet on Obsessive-Compulsive Disorder," pp. 226–227.

16. Raymond and Janet, *Névroses et Idées Fixes,* 2: 385.

17. Janet, *Névroses et Idées Fixes,* 1, quoted in Pitman, "Pierre Janet on Obsessive-Compulsive Disorder," p. 226.

18. Rouart, "Tics et Personnalité de l'Enfant," pp. 267–292, see p. 271.

19. Ibid., p. 292, italics in original.

20. Ibid., pp. 270–273.

21. Ibid., pp. 272–273.

22. Ibid., p. 274.

23. Ibid., p. 276.

24. Ibid., pp. 278–279.

25. Ibid.

26. Rouart would, however, remain active in the psychoanalytic movement.

His 1954 book, *Psychopathologie de la Puberté et de l'Adolescence* (Paris: Presses Universitaires de France) was extemely influential among psychoanalysts. As late as the 1990s Rouart continued to advocate the effectiveness of psychoanalytic therapy. See Julien Rouart, "Psychanalyse et Guerison," *Revue Francaise de Psychanalyse,* 55 (1991): 437–446.

27. Joseph Alliez and Serge Audon, "Reflexions sur la Maladie des Tics de Gilles de la Tourette," *La Provence Médicale,* 12 (1972): 9–15; quotation from p. 10; also see p. 14.

28. The history of the French psychoanalytic movement is complicated by a variety of factors, including but not limited to the Nazi occupation and attitude toward psychoanalysis. See L. Singer, "La Place de la Psychiatrie Française dans la Psychiatrie Européenne," *Annales Médico-Psychologiques,* 151 (1993): 256–259. Also see Alain De Mijolla, "La Psychanalyse et les Psychanalystes en France entre 1939 et 1945," *Revue Internationale d'Histoire de la Psychanalyse,* 1 (1988): 167–223; and Elisabeth Roudinesco, *Jacques Lacan & Co.: A History of Psychoanalysis in France, 1925–1985,* trans. Jeffrey Mehlman (Chicago: University of Chicago Press, 1990), esp. pp. 151–164.

29. This was so, even though French psychoanalysts themselves split into a number of warring factions that resulted in serious schisms in the 1950s.

30. P. F. Girard and Schott, "Le Mécanisme Physio-Pathologique des Tics de la Gaucherie au Tic," *Pédiatrie* (1948): 2–24; quotation from p. 2. Also see P. F. Girard, "Le Rôle de la Gaucherie Corigée dans la Genèse des Tics," *Pédiatrie,* numéro spécial, 1948.

31. Girard and Schott, "Le Mécanisme Physio-Pathologique des Tics de la Gaucherie au Tic," p. 21 (italics in original). Also see p. 22 for another example of the authors' deference toward Rouart. In this instance, Girard and Schott clearly wish to disagree with Rouart, but write, "We are not looking to contest this interpretation, but it is difficult to verify; it has been much easier to prove that the young boy was a contrary lefthander" (p. 22).

32. P. F. Girard, "Les Tics," *Traité de Médecine,* 16 (1949): 1183–1190; quotation from p. 1188. Unfortunately, Girard added, experience had demonstrated that "progressive tics have the tendency to generalize, rebelling against isolation, psychomotor reeducation, and psychotherapy and constitute a grave illness for the patient." Given this, Girard believed that practitioners were justified in using "electro-shock, a new therapeutic whose mechanism of action eludes us, but which has allowed us nevertheless to definitely cure, at least to relieve in an extremely noticeable manner, the most florid tics" (p. 1190).

33. Lebovici, *Les Tics chez l'Enfant.*

34. Roudinesco, *Jacques Lacan & Co.,* pp. 182–186.

35. Lebovici, *Les Tics chez l'Enfant,* pp. 1–6, 131.

36. Lebovici, *Les Tics chez l'Enfant*, pp. 11–31; quotations from pp. 96–98.
37. Ibid., pp. 99–100.
38. Ibid., pp. 105–110.
39. Ibid., pp. 59, 115–116, 131.
40. Ibid., pp. 81–82.
41. Ibid., p. 132.
42. Serge Lebovici, "Les Tics chez l'Enfant: Introduction à la Discussion," *Neuropsychiatrie de l'Enfance et de l'Adolescence,* 31 (1983): 169–170; Serge Lebovici, J.-F. Rabain, Tobie Nathan, R. Thomas, and M.-M. Duboz, "A Propos de la Maladie de Gilles de la Tourette," *Psychiatrie de l'Enfant,* 29 (1986): 5–59.
43. Claude Launay, "Les Tics chez l'Enfant," *Revue du Praticien,* 4 (1954): 2783–2790, see p. 2790; C. Dinard, "Considérations sur la Genèse des Tics chez l'Enfant," *Pédiatrie,* 14 (1959): 778–784.
44. P. Geissmann, A. Lévy, and L. Israël, "Présentation de Film: Un Cas de Syndrome de Gilles de la Tourette," *Annales Médico-Psychologiques,* 118 (1960): 578.
45. Simone Sylvie Bloch, "Contribution à l'Étude de la Maladie de Gilles de la Tourette (À Propos d'un Cas)," Thèse pour le Doctorat en Médecine, no. 60. (Strasbourg: Faculté de Médecine de Strasbourg, 1960) 70 pp.; quotations from pp. 23, 61–62.
46. H. Faure, *Cure de Sommeil Collective et Psychothérapie de Groupe* (Paris. Masson, 1958); H. Faure, M. C. Faure, F. Roux-Filio, G. Veynante, and A. Igert, "Cure de Sommeil Collective avec Psychothéropie en Neuropsychiatrie Infantile," *Annales Médico-Psychologiques,* 118 (1960): 47–82; Paule Aimard and C. Kohler, "Resultats de la Cure de Sommeil Associée à la Psychotherapie dans 10 Cas de Maladie de Gilles de la Tourette," *Revue de Neuropsychiatrie Infantile et d'Hygiène Mentale de l'Enfance,* 9 (1961): 110–122.
47. Thus, reviewing the case of a twenty-four-year-old male with coprolalia, choreiform motor tics, and obsessive behaviors, a team of psychiatrists concluded that the cause was psychogenic because of the man's "homosexual tendencies." P. Mouren, M. Berthon, R. Paché, and J. Alliez, "Maladie des Tics Associée à une Névrose Obsessionnelle Sévère," *Annales Médico-Psychologiques,* 122 (1964): 821.
48. Angelo Taranta and Gene H. Stollerman, "The Relationship of Sydenham's Chorea to Infection with Group A Streptococci," *American Journal of Medicine,* 20 (1956): 170–175.
49. Because of haloperidol's side effects, they switched to the phenothiazine tranquilizer, thioproperazine mesylate, a dopamine suppressor similar to chlorpromazine.
50. M. Langlois and L. Force, "Révision Nosologique et Clinique de la Maladie de Gilles de la Tourette Évoquée par l'Action de Certains

Neuroleptiques sur son Évolution," *Revue Neurologique,* 113 (1965): 641–645.

51. Yves Ranty, "Étiologie et Thérapeutique de la Maladie des Tics de Gilles de la Tourette," Thèse pour le Doctorat en Médecine, Université de Bordeaux, Faculté Mixte de Médecine et Pharmacie, no. 16. (Bordeaux: Éditions Bergeret, 1967), 100 pp.; quotations from pp. 90–93.

52. Ibid., p. 91.

53. Ibid., pp. 92–93.

54. J. M. Léger and Yves Ranty, "Aspects Psychopathologiques du Tic de la Maladie de Gilles de la Tourette," *Annales Médico-Psychologiques,* 2 (1969): 689–695; quotation from pp. 690–694.

55. J. M. Léger and Yves Ranty, "Evolution of Three Cases of Gilles de la Tourette's Syndrome Observed during Four Years," in Faruk Abuzzahab and Floyd Anderson, eds., *Gilles de la Tourette's Syndrome* (St. Paul, Minn.: Mason Publishing, 1976), pp. 123–129; quotation from p. 127.

56. Ibid., p. 692.

57. Yves Ranty, "L'Aspect Psychiatrique de la Maladie des Tics de Gilles de la Tourette," Université de Bordeaux, Faculté Mixte de Médecine et de Pharmacie (1970), p. 54.

58. Ibid., see esp. pp. 33–51.

59. Ibid., see esp. pp. 52–53.

60. Marie Olivennes, "Maladie des Tics," *Annales de Pédiatrie,* 17 (1970): 911–914.

61. "We believe . . . that Gilles de la Tourette's disease is a chronic psychiatric affliction . . . supposing a psychopathological organization (here psychotic) assuredly resting, but not reduced, to a concomitant or preliminary cerebral alteration (here, a biochemical lesion in the striate cortex)." M. Yvonneau and P. Bezard, "Sur un Cas de Maladie des Tics Bloquée par le Sulpiride. Étude Psycho-Biologique," *L'Encephale,* 59 (1970): 439–459; quotations from pp. 453–455.

62. Ibid., p. 454. See Hilde Bruch and Lawrence C. Thum, "Maladie des Tics and Maternal Psychosis," *Journal of Nervous and Mental Disease,* 146 (1968): 446–456. Yvonneau reached similar conclusions in an article published two years later. M. Yvonneau, "Étude Biologique de Deux Cas de Maladie de Gilles de la Tourette," *Revue Neurologique,* 126 (1972): 65–70.

63. Alliez and Audon, "Reflexions sur la Maladie des Tics de Gilles de la Tourette," pp. 9–15; quotations from p. 12, italics in original.

64. Ibid., pp. 10, 15. A similar conclusion was reached by the Swiss/French psychiatrist M. de Morsier, whose 1951 study was discussed earlier. Reporting on eleven patients he had followed for ten to thirty-six years (his first publication on tics was in 1951), de Morsier examined the conflict between the organic and psychoanalytic views and, like Alliez and Audon,

called for a more cooperative approach. G. de Morsier, "La Maladie des Tics—Étude de 11 Cas," *Archives Suisses de Neurologie, Neurochirurgie de Psychiatrie,* 3 (1972): 15–28.

65. Joseph Alliez and Serge Audon, "La Maladie des Tics de Gilles de la Tourette," *Annales Médico-Psychologiques,* 2 (1975): 489–522, see pp. 509–510; quotation from p. 514. They were still certain that there was no infectious etiology.

66. For a history of the conflicts between Lacanian and orthodox psychoanalysts, see Roudinesco, *Jacques Lacan & Co.,* pp. 223–477, and Elisabeth Roudinesco, *Jacques Lacan,* trans. Barbara Bray (New York: Columbia University Press, 1997), pp. 399–427.

67. Lebovici, "Les Tics chez l'Enfant," pp. 169–170; quotation from p. 169.

68. Ibid., p. 170.

69. P. Moran, "Aspects Semniologiques," *Neuropsychiatrie de l'Enfance et de l'Adolescence,* 31 (1983): 170–177, see esp. p. 170.

70. Martine Lefevre and Serge Lebovici, "L'Application de la Nouvelle Classification Americaine Dite DSM III, à la Psychiatrie de l'Enfant et de l'Adolescent," *Psychiatrie de l'Enfant,* 26 (1983): 459–505. Also see Serge Lebovici, "Note à propos de la Classification des Troubles Mentaux en Psychopathologie de l'Enfant," *Psychiatrie de l'Enfant,* 31 (1988): 135–149.

10. The Triumph of the Organic Narrative

1. Claire Gold (a pseudonym), "Something Terrible Was Happening to Our Son," *Good Housekeeping* (September 1976): reprint, 4 pp.; taped interview with "Claire Gold," New York City, 12 May 1997.

2. Ibid.

3. Sally Wendkos Olds, "A Nightmarish Disease: Terrible to Have: Terrible to Live With," *Today's Health* (October 1975), pp. 40–43; quotation from p. 40.

4. Gold, "Something Terrible Was Happening to Our Son"; interview with "Claire Gold," 12 May 1997.

5. Ibid., p. 4.

6. Barry Kramer, "Rare Illness Reduces Its Victims to Shouts, Grunts—and Swearing," *Wall Street Journal,* 20 June 1972, pp. 1, 15.

7. Edward Edelson, "The 'Foulmouth Disease'—It Can Be Cured," *New York Daily News,* 27 December 1972, p. 60.

8. Jane E. Brody, "Bizarre Outbursts of Tourette's Disease Victims Linked to Chemical Disorder in the Brain," *The New York Times,* 29 May 1975, p. 70.

9. For instance, see *The Baltimore Evening Sun,* 16 June 1975, B2.

10. "New Drug Controls Tourette's Syndrome," *Reader's Digest,* March

1973; also see the editors of *Reader's Digest* to Abbey S. Meyers (vice president of TSA), 2 May 1978, Tourette Syndrome Association Archives, Bayside, New York (hereafter cited as TSA archives); Gloria Hochman, "A Mysterious Malady: Involuntary Cursing and Anti-Social Behavior Are Often Symptoms of a Disorder Known as Tics of Tourette's," *Philadelphia Inquirer,* March 1976, p. 1.

11. For instance, see Judy Grande, "Tourette Syndrome," *Sunday Magazine* of the Rockland County (New York) *Journal News,* 15 August 1976; *Newsday,* the large circulation Long Island paper, reprinted Sally Olds's *Today's Health* AMA article in its Sunday magazine, 19 October 1975.

12. Interview with Judy Wertheim, Elaine Novick, Betti Teltscher, and Erica Feinholtz, 12 May 1997, New York City. All were officers of the Tourette Syndrome Association in the 1970s to mid-1980s. The role of the Association in the preparation of these stories, including Sally Olds's 1975 article in *Today's Health,* is also documented in a mailing to members of the Tourette Syndrome Association in October, 1975, p. 2. TSA archives.

13. The original report of this case appeared in Arthur K. Shapiro and Elaine Shapiro, "Treatment of Gilles de la Tourette's Syndrome with Haloperidol," *British Journal of Psychiatry,* 114 (1968): 345–350. A more accessible version is in the introductory chapter of Arthur K. Shapiro, Elaine S. Shapiro, Ruth D. Bruun, and Richard D. Sweet, *Gilles de la Tourette Syndrome* (New York: Raven Press, 1978), pp. 1–9; quotation from p. 1.

14. Shapiro and Shapiro, "Treatment of Gilles de la Tourette's Syndrome with Haloperidol," p. 345.

15. Ibid., pp. 345–350.

16. Elaine Shapiro sees herself as having played a secondary role to Arthur in the work on Tourette syndrome. Arthur "was really doing the major work and I was ancillary, but I wasn't thinking about some of the issues as deeply as he was," Elaine asserted in our 13 June 1996 interview. However, an examination of the record and numerous interviews with those who worked with both researchers during these years suggests that her role ultimately was equal to his. Arthur would discuss every aspect of his work with Elaine and in almost every publication they were joint authors, with Elaine taking the primary authorship of articles on education and testing.

17. Interview with Elaine S. Shapiro, 13 June 1996.

18. "Each drug was repeated two or three times," including placebos, "pentobarbitone, 90 mg., chlorpromazine, 50 mg., meprobamate, 200–800 mg., diazepam 5 to 10 mg.," with no results. "Slight effect on symptoms was noted with diazepam 20–40 mg., imipramine and amitriptyline 25 to 50 mg." The "best" results were achieved with "amitriptyline 75 mg. q.i.d [four times per day or 300 mg] and diazepam 10 mg. q.i.d., but this dosage had to be discontinued because of severe mydriasis" (p. 347). "Im-

provements had been reported for a combination of drugs which were tried on this patient in the following dosages: trimethadione 300 mg. b.i.d. [twice a day] thioridazine 25 to 100 mg. q.i.d., and trifuoperazine 20 to 50 mg. t.i.d. [three times a day]." Responses were good at even higher dosages, but, reported the Shapiros, "drowsiness, akathisia [restlessness, a need to walk or pace, or the inability to sit for a long period] and mydriasis [dilation of the pupil] were severe." Shapiro and Shapiro, "Treatment of Gilles de la Tourette's Syndrome with Haloperidol," p. 347.

19. "Akathisia and akinesia [muscle weakness, fatigue, and, in extreme cases, the inability to move muscles] occurred at dosages over 4 mg. per day, and Parkinsonism occurred at 10 mg. per day," ibid., p. 347.

20. Ibid., p. 347.

21. Shapiro and Shapiro, "Treatment of Gilles de la Tourette's Syndrome with Haloperidol," p. 349.

22. Ibid., p. 349.

23. Arthur K. Shapiro and Elaine Shapiro, "Clinical Dangers of Psychological Theorizing," *Psychiatric Quarterly*, 45 (1971): 159–171.

24. Arthur K. Shapiro, Elaine Shapiro, and Henriette Wayne, "Treatment of Tourette's Syndrome with Haloperidol, Review of 34 Cases," *Archives of General Psychiatry*, 28 (1973): 92–97. Also see Tourette Syndrome Association Newsletter, October 1975, hereafter cited as TSA NL.

25. Arthur Shapiro, interview, 21 September 1994.

26. Because the basal ganglia (particularly the striate cortex—putamen and caudate nucleus) is associated with motor movements, one logical assumption is that the involuntary motor tics and uncontrolled vocalizations are related to hypertransmission or reception of dopamine in the basal ganglia area.

27. Arthur Shapiro, interview, 21 September 1994.

28. Arthur K. Shapiro, "Symposium on Gilles de la Tourette's Syndrome," Panel Roundtable at American Psychiatric Association Meeting, 16 May 1968, published in the *New York State Journal of Medicine* (1 September 1970): 2193–2214.

29. Arthur K. Shapiro, "Dangers of Premature Psychologic Diagnosis," ibid., pp. 2210–2213; quotation from p. 2211. The Shapiros would return to this theme often. See Shapiro and Shapiro, "Clinical Dangers of Psychological Theorizing," pp. 159–171.

30. Shapiro, "Discussion," ibid., p. 2214.

31. Shapiro, Shapiro, and Wayne, "Treatment of Tourette's Syndrome with Haloperidol, Review of 34 Cases," p. 96.

32. Arthur K. Shapiro, Elaine Shapiro, Henriette L. Wayne, John Clarkin, "Organic Factors in Gilles de la Tourette's Syndrome," *British Journal of Psychiatry*, 122 (1973): 659–664; quotation from p. 663. Similar argu-

ments are found in Arthur K. Shapiro, Elaine Shapiro, Henriette L. Wayne, John Clarkin, and Ruth D. Bruun, "Tourette's Syndrome: Summary of Data on 34 Patients," *Psychosomatic Medicine,* 35 (1973): 419–435, p. 433.

33. "Testimonial of a Mother and Son Diagnosed in 1975," reprinted in Arthur K. Shapiro, Elaine S. Shapiro, Ruth D. Bruun, and Richard D. Sweet, *Gilles de la Tourette Syndrome* (New York: Raven Press, 1978), pp. 399–402; quotation from p. 402.

34. "Experience of a Father with a Son with Tourette Syndrome, June 1975, ibid., pp. 402–405; quotations from pp. 402, 405.

35. "Experience of a Mother and Two Children," August 1975, ibid., quotations from pp. 405–408.

36. Betti Teltscher, interview, 14 June 1996.

37. Mr. and Mrs. Martin Levey to Mrs. Schiff (editor of the *New York Post*), 13 December 1970, TSA archives.

38. Kramer, "Rare Illness Reduces Its Victims to Shouts, Grunts—and Swearing," p. 15. The son's name was not actually "Bill."

39. Levey to Schiff, 13 December 1970.

40. Kramer, "Rare Illness Reduces Its Victims to Shouts, Grunts—and Swearing," p. 15. "Bill" had been referred to Arthur Shapiro in 1967 by George Challas in Iowa, who believed that Bill was a good candidate for treatment with haloperidol. Interview with Elaine S. Shapiro, 23 May 1997.

41. Interview with Wertheim, Novick, Teltscher, and Feinholtz, 12 May 1997.

42. Abby Avin Belson, "The Tourette Syndrome Association," published as an appendix to Shapiro et al., *Gilles de la Tourette Syndrome,* pp. 409–411.

43. TSA NL, 4, April 1977.

44. "Gilles: Official Newsletter of the Gilles de la Tourette's Syndrome Association," 1 (Spring 1974), p. 2; TSA NL, 4, April 1977, p. 1.

45. Interview with Sy Goldis, Jericho, New York, 10 October 1997.

46. This one is taken from Public and Commercial Notices, *The New York Times,* 6 October 1974, but it first appeared in 1972. These notices resulted in more than one hundred responses by 1974 (TSA NL, Fall 1974).

47. "Gilles," 1 (Spring 1974), p. 2; TSA NL, 4, April 1977, p. 1.

48. Among those articles were: Shapiro, Shapiro, and Wayne, "Treatment of Tourette's Syndrome with Haloperidol, Review of 34 Cases"; Elaine Shapiro, Arthur K. Shapiro, Richard D. Sweet, and Ruth D. Bruun, "The Diagnosis, Etiology and Treatment of Gilles de la Tourette's Syndrome," in D. V. Siva Sankar, ed., *Mental Health in Children, Volume I: Genetics, Family and Community Studies* (Westbury, N.Y.: PJD Publication, 1975), 167–173; Arthur K. Shapiro and Elaine S. Shapiro, *Tic Syndrome and other Movement Disorders: A Pediatricians Guide,* pamphlet produced for the TSA and funded by the Laura B. Vogler Foundation, Inc., Bayside, New York, 1980.

49. "Gilles," p. 2; TSA NL, 4, April 1977, p. 1.
50. "Gilles" I: 1.
51. TSA NL, Fall 1974. For the early results of this international registry see Faruk E. Abuzzahab and Floyd O. Anderson, "Gilles de la Tourette's Syndrome," *Minnesota Medicine*, 56 (1973): 492–496. Abuzzahab and Anderson reported that they had conducted "an exhaustive search of the literature" and set up "an international registry . . . based upon information consolidated from 430 published cases plus 55 case reports that have come to our attention through the international registry" (p. 496).
52. Sheldon Novick, "Reaching Physicians, Encouraging Research: The Role of the Medical Director," TSA NL, Summer 1982, pp. 13–14.
53. B. Teltscher, interview, 14 June 1996, New York City. Aside from Sacks, Novick was responsible for recruiting a number of prominent researchers including Eldridge and British psychiatrist J. A. Corbett. See correspondence of Sheldon Novick, 1974–1983, private collection in possession of Elaine Novick, New York City. Hereafter cited as Novick correspondence.
54. TSA NL, 4, April 1977, p. 1.
55. Wertheim, interview, 12 May 1997; Brody, "Bizarre Outbursts of Tourette's Disease Victims Linked to Chemical Disorder in the Brain," p. 70.
56. TSA NL, 1, Spring 1974; TSA NL, Fall 1974; Letter to members, 23 October 1974, TSA archives.
57. TSA NL, January 1975; Oliver W. Sacks to Sheldon Novick, 25 July 1974, London, England. Novick correspondence.
58. TSA NL, January 1975. This was based on the belief that increasing dopamine transmission would flood brain receptors and eventually regulate or "normalize" the transmission and reception of dopamine.
59. Elaine Shapiro et al., "Diagnosis, Etiology and Treatment of Gilles de la Tourette's Syndrome."
60. Sheldon Novick attempted to interest a number of prominent psychoanalysts to hold sessions on Tourette's at psychiatric meetings, but his correspondence reveals a lack of interest from this community. See Peter B. Neubauer to Novick, 16 July 1976; Novick to Margaret Mahler, 21 July 1975; Novick to Z. Alexander Aarons, 19 August 1975.
61. Interview with Feinholtz, Novick, Teltscher, and Wertheim, 12 May 1997. Guthrie was enormously helpful to the organization in its early years. See TSA NL, January 1976, p. 1.
62. Including Friedhoff's neurotransmitter study; Van Woert's serotonin study; Sacks's neurological and emotional family study; Jose Yaryura-Tobias's (director of research at North Nassau Mental Health Center) chlorimipramine trial for patients who did not respond to haloperidol; and Miller's "new medication trial."
63. See letter "Dear Doctor" from Sheldon Novick, 2 May 1975, TSA archives.

64. TSA archives.
65. *Saturday Review,* 3 April 1976, p. 21 (italics in original); *U.S. News & World Report,* 19 July 1976; *Medical Economics,* 20 September 1976. The ad also appeared in the *Madison Avenue Magazine* in June 1976 and *Southern Living Magazine* in November 1976.
66. TSA NL, July 1976, p. 3.
67. The staff continued to solicit stories about the disorder as well as to provide materials for potentially interested writers, newspapers, and magazines. For example, see Joseph P. Coogan (ed. *Medicine and Mind* and *Today in Psychiatry*) to Eleanor Pearl, 27 May 1976; Pearl to Coogan, 2 June 1976, TSA archives. Also see the *Tampa Tribune,* 19 May 1976; TSA NL, July 1976, p. 3.
68. David R. Zimmerman, "Medicine Today," *Ladies' Home Journal* (April 1975), p. 60; TSA NL, 16 April 1975.
69. "New Drug Controls Tourette's Syndrome," *Reader's Digest* (March 1973); "Cure for Foul Mouths," *Science Digest* (September 1973), p. 53; Daniel Goleman and Jerome Engel, Jr., "A Feeling of Falling," *Psychology Today* (November 1976), pp. 107–108.
70. Anita Ricterman, "Problem Line," *Newsday,* 1 August 1975; T. R. Van Dellen, M.D., "The Family Doctor," *New York Daily News,* 10 December 1976. Also see Dr. Lester Coleman, "Speaking of Your Health," (syndicated) 29 September 1976 cited in TSA NL, October 1976, p. 4.
71. TSA NL, October 1975.
72. Shapiro, Shapiro, and Wayne, "Treatment of Tourette's Syndrome with Haloperidol, Review of 34 Cases." Also see TSA NL, October 1975.
73. Letter from Cy Goldis to "Friends," 18 February 1975, TSA archives.
74. In 1976 Abuzzahab and his collaborator and resident, Floyd O. Anderson, published a collection of papers presented (or invited to be presented) at a panel at the 1970 American Psychiatric Association meeting in San Francisco. The panel was titled the "International Registry of Gilles de la Tourette's Syndrome." Also included in the volume was the editors' summary account on "current knowledge" about Tourette's. Many of the papers focused on the extent to which the actions of haloperidol, in Abuzzahab's words, "open the door to the possibility of verifying the neurochemical lesion of the disorder" (p. ix). The collection is extremely interesting because it includes essays from clinicians and researchers throughout Europe and one from a researcher in China. The contributors vary greatly in their assessment about many matters, not least of all whether or not the clinical and pharmacological evidence supports Abuzzahab's own point of view. The editors planned that this would be the first of three volumes. The second would provide English-language translations of the key nineteenth-century French works and the third would publish papers on Tourette syndrome presented at the World Congress

of Psychiatry in 1977. Unfortunately, neither of these appeared. Faruk S. Abuzzahab and Floyd O. Anderson, eds., *Gilles de la Tourette's Syndrome* (St. Paul: Mason Publishing Co., 1976).

75. TSA NL, July 1976, pp. 1, 4–5. Letter from Elaine Novick to TSA members and "Friends," 24 June 1975, TSA archives.

76. TSA NL, July 1976, pp. 1–2; ibid., October 1976, pp. 3–4.

77. TSA NL, 7 (2), April 1980, p. 2; interview with Abbey Meyers, 21 May 1997. A few months later, Meyers was hired as TSA coordinator.

78. TSA NL, 9 (2), Spring 1982, pp. 1–3.

79. TSA NL, 8 (2), April 1981, p. 2.

80. TSA NL, 8 (1), January 1981, p. 6.

81. TSA NL, 9 (2), Spring 1982, pp. 1–3.

82. Shapiro and Shapiro, *Tic Syndrome and Other Movement Disorders: A Pediatricians Guide*, pp. 8, 14. The Shapiros wrote that the "evidence includes the following findings: Organic abnormalities are found significantly more frequently in patients with Tourette syndrome than controls: 48.8% with electroencephalograph abnormalities (compared to 5–15% expected in the population); 68% with organic findings on psychological tests (compared to 28% in controls); 57% with soft neurological signs (compared to an expected 20% in children); left-handedness or ambidexterity in 23% (compared to 5–10% in the population); 68% with minimal brain dysfunction (compared to the expected 3–7%); 3 to 1 male to female ratio (similar to the ratio in other organic conditions such as dystonia musculorum deformans, autism, minimal brain dysfunction); 34.5% with a history of tics in primary family members (compared to an expected percentage of 12–24%); genetic factors: 40 families with more than one first-degree relative with Tourette syndrome, 7 of 9 identical twin pairs concordant for Tourette syndrome, and 4 of 5 fraternal twins discordant for Tourette syndrome."

83. The bill provided subsidies for liability insurance for users of these drugs, support for researchers, and funds for studies to determine future needs for orphan drugs. Also, if no private company could be persuaded to produce these drugs, a federal agency created by the legislation was authorized to do so. TSA NL, 8 (2), April 1982, p. 2; TSA NL, 9 (2), Spring 1982, p. 2; TSA NL, 9 (3), Summer 1981, pp. 6–7; TSA NL, Spring 1983, p. 9. The hearings also helped to launch a career for Abbey Meyers, who is executive director of NORD, the National Organization for Rare Disorders.

84. TSA NL, 9 (2), Spring 1982, p. 8–9; ibid., 9 (3) Summer 1982, pp. 11–12.

85. TSA, "Special Report on the First International Gilles de la Tourette Syndrome Symposium," New York, 27–29 May 1981," TSA archives.

86. Judy Wertheim, "A 10th Anniversary Message. Who We Are, How We Help," TSA NL, 9 (2), Spring 1982, pp. 1–2; quotation from p. 2.

87. Arthur K. Shapiro, "Remarks at the Tenth Anniversary Membership Meeting of the Tourette Syndrome Association," 22 May 1982, Mt. Sinai Hospital, New York City. Reprinted in TSA NL, 9 (3), Summer 1982, pp. 1–2.

88. See, for instance, "Experience of a Mother and Two Children," August 1975, in Shapiro et al., *Gilles de la Tourette Syndrome,* pp. 405–408, esp. pp. 407–408.

89. For instance see J. A. Corbett, Gilles de la Tourette's Syndrome, in Abuzzahab and Anderson, eds. *Gilles de la Tourette's Syndrome,* pp. 3–5; Alexander R. Lucas, "Follow-up of Tic Syndrome," ibid., pp. 13–17.

90. Solomon H. Snyder, *Drugs and the Brain* (New York: Scientific American Library, 1996), pp. 77–89.

91. "Special Report on the First International Gilles de la Tourette Syndrome Symposium."

92. Richard L. Borison, Lolita Ang, Bruce I. Diamond, and John M. Davis, "Is Haloperidol A Specific Agent for Treating Tourette Syndrome?" *Neurology* (April 1982), abstract in TSA NL, 9, Summer 1982, p. 12. Also see Richard L. Borison, Lolita Ang, W. J. Hamilton, Bruce I. Diamond, and John M. Davis, "Treatment Approaches in Gilles de la Tourette's Syndrome," *Brain Research Bulletin,* 11 (1983): 205–208.

93. Shapiro, Shapiro, and Wayne, "Treatment of Tourette's Syndrome with Haloperidol, Review of 34 Cases," p. 96; Shapiro, Shapiro, Bruun, and Sweet, *Gilles de la Tourette Syndrome,* pp. 315–335.

94. Richard L. Jenkins and Barry N. Fine, "Experience in 20 Cases of Tourette's Syndrome," Abuzzahab and Anderson, eds., *Gilles de la Tourette's Syndrome,* pp. 46–67; quotations from p. 48.

95. S. J. M. Fernando, "Gilles de la Tourette's Syndrome: A Follow-up of Four Cases and a New Case Report," in Abuzzahab and Anderson, eds., *Gilles de la Tourette's Syndrome,* pp. 7–12; quotations from pp. 11–12. Also see S. J. M. Fernando, "Gilles de la Tourette's Syndrome: A Report on Four Cases and a Review of Published Case Reports," *British Journal of Psychiatry,* 113 (1967): 607–617.

96. Shapiro quoted by Kramer, "Rare Illness Reduces Its Victims to Shouts, Grunts—and Swearing," p. 5.

97. Ibid.

98. Abuzzahab's first report on his use of haloperidol was in 1971. See Faruk S. Abuzzahab and James K. Ehlen, "The Clinical Picture and Management of Gilles de la Tourette's Syndrome," *Child Psychiatry and Human Development,* 2 (1971): 14–25.

99. Kramer, "Rare Illness Reduces Its Victims to Shouts, Grunts—and Swearing," p. 5.

100. John Teltscher, interview, 14 May 1997, New York City. Despite the fact that Teltscher stopped taking haloperiodol at age twelve, he believes that the side effects adversely affected him well into adolescence.

101. Janice R. Stevens and Paul H. Blachly, "Successful Treatment of the Maladie des Tics, Gilles de la Tourette's Syndrome," *American Journal of Diseases in Children,* 112 (1966): 541–545.

102. "I have treated but four patients with Tourette's syndrome. The first was the subject of a paper by Dr. Janice Stevens. . . Two others were children and I lost track of them several years ago." Paul H. Blachly to Sheldon Novick, 7 April 1975, Novick correspondence.

103. Interview with Abbey Meyers, 21 May 1997. Sy Goldis, an early president of the TSA, remembers a parallel experience for his daughter, also a Shapiro patient. Interview with Sy Goldis, 10 October 1997.

104. Claire Gold to author, 19 June 1997; Mrs. Gold's letter went on: " At the time, Dr. Cohen did not believe that Tommy's school phobia was due to Haldol, but rather to family dynamics. However, as Dr. Cohen began to treat more and more children like Tommy, he recognized the pattern of higher doses of Haldol causing school phobia in some children, and subsequently wrote a paper on it. Fortunately for Tommy, there is a happy ending to his story. Much to Dr. Cohen's objection, we allowed Tommy to resume Haldol (1/2 mg) once he returned home, and to monitor his own dose, under our supervision. We explained to Tommy that there appeared to be a narrow window, within which his symptoms would respond to Haldol, and beyond that, his symptoms would increase, rather than decrease, when he raised the medication. By the time Tommy was 14, he gradually weaned himself off Haldol and has not needed any medication since."

105. Peggy Harmon to TSA, 15 November 1976, TSA NL, 4 January 1977, p. 4.

106. Abbey Meyers, "Education and the Tourette Child," TSA NL, 7 (3), July 1980, p. 2.

107. When the "panelists were asked about the value of psychotherapy in accepting the diagnosis of T. S., Mark [a twenty-three year old] felt it was important to have a good friend who is unbiased. Another panelist felt that it is important to have someone around to talk to and help build up your confidence." "Discussion of Tourette Syndrome in Young Adulthood," TSA NL, 8 (1), January 1981, pp. 1–2.

108. Sheldon Novick, "The Role of the Medical Director of the Tourette Syndrome Association," draft ms. prepared for presentation at the First International Symposium on Gilles de la Tourette's Syndrome and Related Dysfunctions of the Central Nervous System, New York City, 28 May 1981. Novick sent this draft to Arnold Friedhoff on 20 May 1981 for comments. Original in TSA archives.

109. Ruth D. Bruun, "Side Effects and Haloperidol," TSA NL, Spring 1983, p. 5.

110. Arthur Shapiro, interview, 21 September 1994; Elaine Shapiro, interview, 13 June 1996, 13 May 1997.

111. "Interview with Dr. Arthur Shapiro, re: treatment with Haloperidol," TSA NL, 3 (3), July 1976, p. 4; Shapiro et al., *Gilles de la Tourette Syndrome,* pp. 326–335; Shapiro and Shapiro, *Tic Syndrome and Other Movement Disorders: A Pediatricians Guide,* pp. 9–12.
112. Elaine Shapiro, communication, 7 April 1998; interview, 13 June 1996.
113. Borison et al., "Is Haloperidol a Specific Agent for Treating Tourette Syndrome?" TSA NL, 9, Summer 1982, p. 12.

11. Clashing Cultural Conceptions

1. Georges Gilles de la Tourette, "Étude sur une Affection Nerveuse Caractérisée par de l'Incoordination Motrice Accompangnée d'Écholalie et de Coprolalie (Jumping, Latah, Myriachit)," *Archives de Neurologie,* 9 (1885): 19–42, 158–200; Programme du Centenaire du Syndrome de Gilles de la Tourette, Hôpital de la Salpêtrière, Paris, 2–3 May 1985.
2. In the introduction to his 1976 collection of essays Abuzzahab wrote, "thoughts are being devoted to a Centennial Celebration of the publication of Gilles de la Tourette, to be held at the Salpêtrière, Paris, France, 1985." Faruk S. Abuzzahab and Floyd O. Anderson, eds., *Gilles de la Tourette's Syndrome* (St. Paul: Mason Publishing Co., 1976), p. x.
3. "Paris Meeting to Mark Tourette Centennial," TSA NL, 12, Winter 1985, p. 1; ibid., 12, Spring 1985, pp. 1–2.
4. Dugas had corresponded with Sheldon Novick, TSA medical coordinator in the 1970s. See M. Dugas to Novick, Paris, 21 December 1977, Novick correspondence.
5. That year he had published a paper similar to one that he presented at the centenary meeting. See Michel Dugas, "Le Syndrome de Gilles de la Tourette: État Actual de la Maladie des Tics," *La Presse Médicale,* 14 (1985): 589–593.
6. M. Gonce and M. Dugas, "Tourette Syndrome versus Minor Tics: Clinical and Therapeutical Differences," paper presented at the First International Gilles de la Tourette's Syndrome Symposium, New York, 1981. Gonce had been in contact with the Tourette Syndrome Association and the Canadian Tourette Syndrome Foundation since the mid-1970s. See Michel Gonce and André Barbeau to H. E. Steinberg, Montréal, 5 July 1976, Novick correspondence.
7. "Centenaire du Syndrome de Gilles de la Tourette," *Revue Neurologique,* 142 (1986): 810–866.
8. P. Castaigne, "Introduction [Centenaire du Syndrome de Gilles de la Tourette]," *Revue Neurologique,* 142 (1986): 801–802; quotations from p. 801.
9. Two other presentations focused on history. Neurology Professor Gérard Dordain (Clermont-Ferrand) examined the concept of tics from the four-

teenth century until the early twentieth century. Arguing that there had been no historical agreement on what constituted the disorder, Dordain pointed out that Gilles de la Tourette's 1885 publication had not resulted in a unified view. Dordain reminded the audience that "in the two years following [publication] Guinon (1886–1887), Charcot's intern, contested with an astonishing ferocity Gilles de la Tourette's description." Gérard Dordain, "Le Concept de Tic dans l'histoire des Mouvements Anormaux," *Revue Neurologique,* 142 (1986): 803–807; quotation from p. 807. British neurologist A. J. Lees (London) then offered a biographical account, revealing, among other interesting information, that Gilles de la Tourette had been extremely unpopular with his contemporaries at the Salpêtrière. A. J. Lees, "Georges Gilles de la Tourette: The Man and His Times," *Revue Neurologique,* 142 (1986): 808–816.

10. A rereading of Gilles de la Tourette's nine cases, Dugas believed, showed that six of them were consistent with the DSM-III (1980) definition of Tourette syndrome. Michel Dugas, "La Maladie des Tics: D'Itard aux Neuroleptiques," *Revue Neurologique,* 142 (1986): 817–823.

11. Ibid., p. 822. I also relied on an English-language translation of this paper that Professor Dugas graciously gave me. The previous year Dugas had reviewed the different biological explanations for these symptoms, concluding that "it was not possible today to propose a coherent biochemical explanation of these symptoms. Does the diversity of results of metabolic and pharmacological studies signify that the syndrome is heterogeneous or rather does it simply reflect the variety of ways by which the central nervous system of individuals who have Gilles de la Tourette's syndrome respond to a still unknown underlying alteration of the brain." Michel Dugas, "Le Syndrome de Gilles de la Tourette: État Actual de la Maladie des Tics," *La Presse Médicale,* 14 (1985): 589–593; quotation from p. 592.

12. Elaine Shapiro and Arthur K. Shapiro, "Semiology, Nosology and Criteria for Tic Disorders," *Revue Neurologique,* 142 (1986): 824–832; quotations from pp. 829–830.

13. Rosewell Eldridge and Martha B. Denkla, "Gilles de la Tourette's Syndrome: Etiologic Considerations," *Revue Neurologique,* 142 (1986): 833–839; quotations from p. 839. They also rejected the notion that effectiveness of haloperidol proved that Tourette syndrome was a result of "excess" dopamine in favor of the view of hypersensitivity of dopamine receptors (p. 835).

14. EMG (electromyographic) shows that "simple tics are short bursts," while "complex tics consist of prolonged EMG activity . . . EEG [electroencephalgram] recording has revealed abnormalities in about 25 to 75 p.[er] 100 of Gilles de la Tourette's patients, but these findings lack specificity." M. Gonce, "Du Movement au Tic: Aspects Neurologiques," *Revue Neurologique,* 142 (1986): 845–850; quotations from p. 845.

15. This also "suggests hypofunction in some striatal and corticolimbic area." T. N. Chase, V. Geoffrey, M. Gillespie, and G. H. Burrows, "Structural and Functional Studies of Gilles de la Tourette Syndrome," *Revue Neurologique,* 142 (1986): 851–855.

16. The success of antipsychotic drugs in treating TS, they noted, had strengthened the hypothesis that it results from a dopanergic over-activity. But, according to Fog and Regeur, this was probably misleading, because the most effective neuroleptics seem to have a "high affinity for striatal D(2)-receptors." Other drugs that act more generally on dopamine (cholinergic, GABAnergic, and peptidergic) have poor clinical results. R. Fog and L. Regeur, "Neuropharmacology of Tics," *Revue Neurologique,* 142 (1986): 856–859; quotation from p. 856.

17. Arthur J. Friedhoff, "Insights into the Pathophysiology and Pathogenesis of Gilles de la Tourette Syndrome," *Revue Neurologique,* 142 (1986): 860–864; quotation from p. 860.

18. Elisabeth Roudinesco, *Jacques Lacan & Co.: A History of Psychoanalysis in France, 1925–1985,* trans. Jeffrey Mehlman (Chicago: University of Chicago Press, 1990), pp. 357–358.

19. Widlöcher was one of the twelve psychoanalysts, along with Lebovici and Lacan, listed in *Who's Who in France, 1982–1983.* Ibid., p. 707.

20. Daniel Widlöcher, "Phénoménologie des Tics: Approche Psychologique et Psychiatrique," *Revue Neurologique,* 142 (1986): 840–844; quotation from pp. 840–841.

21. J. A. Obeso, J. C. Rothwell, and C. D. Marsden, "Simple Tics in Gilles de la Tourette's Syndrome are Not Prefaced by a Normal EEG Potential," *Journal of Neurology and Neurosurgical Psychiatry,*" 44 (1981): 735–738.

22. Widlöcher was not so much endorsing this explanation as pointing out the extent to which psychogenic and neurological issues could not be separated in explorations of the etiology of tics.

23. Widlöcher, "Phénoménologie des Tics," p. 842, citing Fathy S. Messiha and J. C. Carlson, "Behavioral and Clinical Profiles of Tourette's Disease: A Comprehensive Overview," *Brain Research Bulletin,* 11 (1983): 195–204.

24. Ibid., p. 844.

25. MacDonald Critchley, "What's in a Name," *Revue Neurologique,* 142 (1986): 856–866.

26. Telephone interview of 16 June 1997.

27. Serge Lebovici, J.-F. Rabain, Tobie Nathan, R. Thomas, and M.-M. Duboz, "A Propos de la Maladie de Gilles de la Tourette," *Psychiatrie de l'Enfant,* 29 (1986): 5–59.

28. Ibid., quotations from pp. 6, 8–9.

29. Ibid., p. 6. "(1) Multiple tics . . . are obsessive; but, in a certain case, they

coincide with obsessions or have all psychopathological characteristics. (2) They are ameliorated . . . by certain neuroleptic medications which act on the neurotransmitter system. (3) It is thus possible to act on obsessional symptoms—without tics—and on those of obsessional psychoneurosis thanks to these medications. (4) That is to say that the symptoms of obsessional neurosis are of cerebral origin."

30. Ibid., pp. 7, 14–15.

31. Ibid., pp. 15–21, 25, 27, 29.

32. Ibid., p. 30.

33. Abdelaziz had been placed on a regimen of haloperidol, which he often had refused to take.

34. Ibid., p. 31. The therapists reported that they faced several difficulties in treating Abdelaziz, including "on the one hand, the intense inhibition of the patient and . . . on the other hand, competition . . . with the cultural familial traditions," including the continuing consultations with traditional religious therapists. For more on Nathan's therapeutic approach and views on Tourette syndrome, see Tobie Nathan, *L'Influence Qui Guérit* (Paris: Éditions Odile Jacob, 1994) pt. 3: pp. 223–332. Also see Tobie Nathan, *Psychanalyse Païenne: Essais Ethnopsychanalytiques, Nouvelle Édition* (Paris: Odile Jacob, 1995), esp. pp. 109–113.

35. Lebovici et al., "A Propos de la Maladie de Gilles de la Tourette," pp. 32–33.

36. Ibid., p. 47.

37. Ibid., pp. 48, 54, 57–58.

38. Ibid., pp. 57–58.

39. Elaine Shapiro, interview of 13 June 1996. She also expressed similar views expressed in an interview on 13 May 1997.

40. "I think it best to see patients in the more personal atmosphere of my own small office here and not to entangle what will be essentially a personal and voluntary 'investigation' (wrong word!—much too impersonal) with my basic day-to-day work at the hospital," Sacks had written in response to the Tourette Association's 1975 offer to refer patients to him for a proposed clinical investigation. "My present feeling," he explained, "is that *at the start*—whatever may eventuate later—I would do best to see patients (and their families) *alone,* in my own way, and with my own unaided senses of judgement . . . *Later,* perhaps, one might utilize more sophisticated methods of observation and bring in other colleagues." Oliver W. Sacks to Sheldon Novick, 25 July 1974, Novick correspondence (italics in original). Also see TSA NL, January 1975 for a report that Sacks had begun "intensive clinical studies of Tourette patients and their families."

41. Oliver Sacks, *The Man Who Mistook His Wife for a Hat* (New York: Harper and Row, 1987), pp. 92–101.

42. Oliver Sacks, "A Surgeon's Life," *The New Yorker* (March 16, 1992), pp. 85–94. Reprinted in Oliver Sacks, *An Anthropologist on Mars* (New York: Knopf, 1995).

43. Interview with Arthur K. and Elaine S. Shapiro, 30 September, 1994.

44. Elaine Shapiro, interview, 13 May 1997. Also see Arnold Friedhoff, taped interview, 9 October 1997, New York City.

45. "He was convinced, and had always been convinced," Elaine Shapiro told me, "that there is so much similarity between some of the symptoms of Tourette's and it's always hard to know whether you're seeing a tic or whether you're seeing a compulsion. And to the very end, he felt that it was a question really of doing a very fine diagnosis to make that differentiation. He felt that a lot of people just weren't doing it. The literature had just picked up the obsessive-compulsive connection and flown with it, and because of that, the degree of careful scrutiny of the symptoms just wasn't being developed. That's the way he felt. People weren't questioning the diagnosis. He thought it was wrong to call what we saw as tics, compulsions." Interview of 13 May 1996.

46. Interview with Arthur K. and Elaine S. Shapiro, 30 September 1994.

47. Arthur K. Shapiro and Elaine S. Shapiro, "Evaluation of the Reported Association of Obsessive-Compulsive Symptoms or Disorder with Tourette's Disorder," *Comprehensive Psychiatry*, 33 (1992): 152–165.

48. Taped interview with Ruth D. Bruun, 15 May 1996, Great Neck, New York.

49. Elaine Shapiro, interview, 13 June 1996.

50. Although a detailed discussion of the possible genetics of TS is not possible in this book, Comings recently has provided an extremely interesting and balanced discussion of the evolution of his own thinking and research, as well as that of his critics. See David E. Comings, *Search for the Tourette Syndrome and Human Behavior Genes* (Duarte, Calif: Hope Press, 1996).

51. Elaine Shapiro, interview, 13 June 1996.

52. J.-P. Guéguen, "Le Syndrome de Gilles de la Tourette (Maladie des Tics)," *Soins Psychiatrie*, 110/111 (1989–1990): 25–27; quotation from p. 27.

53. Which did not stop some psychoanalysts, especially Jacques Lacan, from criticizing Lebovici for his refusal to support Lacan's revisionist Freudianism. See Roudinesco, *Jacques Lacan & Co.*, pp. 237–238.

54. Henry Meige, *Le Juif-Errant à la Salpêtrière*, edited with an introduction by Lucien Israël (Paris: Collection Grands Textes, Nouvelle Objet, 1993); see introduction, pp. 1–19.

55. Henry Meige, "Le Juif-Errant à la Salpêtrière," *Nouvelle Iconographie de la Salpêtrière* (in three parts), 6 (1893): 191–204, 277–291, 333–358.

56. "Pendant que nous y sommes, rappelons la curieuse évolution de la 'maladie' des tics. Salpêtrière dont Meige était le secrétaire de rédaction?

Cette maladie de Gilles de la Tourette, qui n'avait guère jusqu'à ces dernières années qu'un intérêt historique, a connu véritable résurrection aux États-Unis, sous l'impulsion de Sacks (auteur de 'L'homme qui prenait sa femme pour un chapeau'). Fin clinicien et homme d'une grande générosité, qui, ayant eu l'attention attiré par ce syndrome, en découvrit trois en une seule journée, rien qu'en se promenant dans les rues de New-York. Sur quoi il fonda l'"Association Américaine des Tourette,' comme s'il s'agissait d'alcooliques, d'obèses ou d'émotionnels anonymes. On voit d'ici les 'juifs errants de Meige' ou mieux encore de Meige et Charcot. Faire de la misère humaine une maladie, ça ne vous rappelle rien? Le Führer et le Duce, sans compter les autres caudillos et pinochets étaient de paranoïaques. Tous les autres sont bien portant . . . Les racistes sont des malades." Israël, introduction to *Le Juif-Errant à la Salpêtrière* (1993), pp. 5–6.

57. The woman, Bridget Haardt, reports that although her child had classic symptoms, French physicians refused to make a diagnosis of Tourette's. Haardt finally took her son back to the United States where he was quickly diagnosed. Bridget Haardt to Chris Melbye, 7 February 1997; Sue Levi-Pearl to Haardt, 26 March 1997. Tourette Syndrome Association Archives, Bayside, New York.

58. Pierrick Hordé, *Nouvelles Histoires Incroyable de la Médecine* (Paris: Éditions Filipacchi, 1994). See his discussion of Tourette syndrome on pp. 55–73.

59. Cyrille Koupernik to author, 28 August 1995, letter in author's possession.

60. Sophie Caloone, "Le Syndrome de Gilles de la Tourette et les Troubles Obsessionels Compulsifs. Étude de Comorbidité à Propos de 4 Cas," Thèse pour le Diplôme d'État de Docteur en Médecine présentée sous le forme de Memoire en vue de l'Obtention du D.E.S. de Psychiatrie. (Rennes: Université de Rennes, 1995), p. 168.

61. Ibid., p. 6.

62. Ibid., p. 165.

63. Such an essentialist reading of Gilles de la Tourette seems a bit rigid, especially since, as we saw in Chapter 2, Gilles de la Tourette's phrase was taken literally from Jean Itard's 1825 text and was not based on Gilles de la Tourette's actual clinical examination of the marquise. On the other hand, Caloone ignored Georges Guinon's criticism that Gilles de la Tourette ignored his patients' obsessive symptoms.

64. Ibid., pp. 165–167.

65. M. Dobler-Thierry, "Tics, Obsessions et Compulsions: A Propos des Rapports entre: Syndrome de Gilles de la Tourette et Trouble Obsessionnel-Compulsif," Memoire pour le DES de Psychiatrie, Faculté de Médecine, Paris-Sud, 1995; quotation from p. 153.

12. Clinical Lessons

1. Georges Gilles de la Tourette, "Étude sur une Affection Nerveuse Caractérisée par de l'Incoordination Motrice Accompagnée d'Écholalie et de Coprolalie (Jumping, Latah, Myriachit)," *Archives de Neurologie,* 9 (1885): 19–42, 158–200, see p. 200. Also see Théodule Ribot, *The Diseases of the Will,* trans. Merwin-Marie Snell (Chicago: Open Court Publishing, 1915), pp. 56–57.

2. Georges Guinon, "Sur la Maladie des Tics Convulsifs," *Revue de Médecine,* 6 (1886): 50–80; "Tics Convulsifs et Hystérie," *Revue de Médecine,* 7 (1887): 509–519; "Tics Convulsifs," *Dictionnaire Encyclopédique des Sciences Médicales* (Paris: Asselin et Houzeau, 1887), pp. 555–588; Édouard Brissaud, "La Chorée Variable des Dégénérés," *Revue Neurologique,* 4 (1896): 417–431, esp. pp. 428–431. Also see Édouard Brissaud, "La Chorée Variable des Dégénérés," in *Leçons sur les Maladies Nerveuses,* ed. Henry Meige, 2d ser. (Paris: Masson et Companie, 1899), pp. 516–541.

3. Jean-Martin Charcot, *Leçons du Mardi à la Salpêtrière Policliniques, 1887–1888,* 17 January 1888, p. 129; Ibid., 24 January 1888, pp. 152–153; Ibid., 21 February 1888, pp. 209–213; Jean-Martin Charcot, "Des Tics et des Tiqueurs," *La Tribune Médicale,* 25 November 1888, 19: 571–573. Jacques Catrou, *Étude sur la Maladie des Tics Convulsifs (Jumping—Latah—Myriachit),* Thèse pour le Doctorat en Médecine, no. 129 (Paris: Faculté de Médecine de Paris, Henri Jouve, 1890), pp. 40–45.

4. Howard I. Kushner, "Freud and the Diagnosis of Gilles de la Tourette's Illness," *History of Psychiatry,* 9 (1998): 1–25.

5. Henry Meige and E. Feindel, *Tics and their Treatment,* with a preface by Professor Brissaud [revised and updated version of *Les Tics et leur Traitement* (1902)], trans. and ed. S. A. K. Wilson (New York: William Wood and Co., 1907), p. 225.

6. Sandor Ferenczi, "Psycho-Analytical Observation on Tic," *International Journal of Psycho-Analysis,* 2 (1921): 1–30.

7. Erwin Straus, "Über die organische Natur der Tics und der Koprolalie," *Zentralblatt für die gesamte Neurologie und Psychiatrie,* 47 (1927): 698–699.

8. Arthur K. Shapiro and Elaine Shapiro, "Treatment of Gilles de la Tourette's Syndrome with Haloperidol," *British Journal of Psychiatry,* 114 (1968): 345–350.

9. Howard I. Kushner and Louise S. Kiessling, "The Controversy over the Classification of Gilles de la Tourette's Syndrome, 1800–1995," *Perspectives in Biology and Medicine,* 39 (1996): 409–435.

10. Jean-Martin Charcot, *Leçons du Mardi à la Salpêtrière Policliniques, 1887–1888* [Notes de cours de M. M. Blin, Charcot et Colin], handwrit-

ten and printed (Paris: Bureaux du Progrès Médical, 1887), 6 December 1887, p. 38; Jacques Gasser, "Jean-Marin Charcot (1825–1893) et le Système Nerveux, Étude de la Motricité, du Langage, de la Mémoire et d'Hystérie à la Fin du XIXè Siècle" (Ph.D. diss., Ecole des Hautes Études en Sciences Sociales, 1990), p. 205.

11. George A. Brown, "Chorea: Its Relation to Rheumatism and Treatment," *Montreal Medical Journal*, 19 (1890): 581–588; W. M. Donald, "The Relation between Chorea and Rheumatism," *Physician and Surgeon*, 14 (1892): 535–538; Vincent Pagliano, "Rhumatisme et Chorée," *Marseille Médicale*, 28 (1891): 329–334; J. W. Shaw, "Relation Between Chorea and Rheumatism," *The Dominion Medical Monthly and Ontario Medical Journal*, 6 (1896): 145–147; American Pediatric Society, "Relation of Rheumatism and Chorea," *Archives of Pediatrics*, 10 (1893): 1–27; J. C. Wilson, "A Case of Tic Convulsif," *Archives of Pediatrics*, 14 (1897): 881–887.

12. Henry P. Cooke, "The Relation of Chorea to Rheumatism," *Transactions of the Texas State Medical Association* (1892): 146–153; G. Tully Vaughan, "Case of Chorea, Due Probably to Rheumatism and Endocarditis," *Virginia Medical Monthly*, 9 (1882): 669; E. Leredde, "Note sur un Cas d'Endocardite Choréique d'Origine Microbienne Probable," *Revue Mensuelle des Maladies de l'Enfance*, 9 (1891). 217–220; Robert Massalongo, "Chorée Chez Deux Cardiques," *Revue Neurologique*, 3 (1895): 610–615.

13. Angelo Taranta and Gene H. Stollerman, "The Relationship of Sydenham's Chorea to Infection with Group A Streptococci," *American Journal of Medicine*, 20 (1956): 170–175.

14. Gunnar Husby, Ivo van de Rijn, J. B. Zabriskie, Z. H. Abdin, and R. C. Williams, "Antibodies Reacting with Cytoplasm of Subthalamic and Caudate Nuclei Neurons in Chorea and Rheumatic Fever," *Journal of Experimental Medicine*, 144 (1976): 1094–1110.

15. Kushner and Kiessling, "The Controversy over the Classification of Gilles de la Tourette's Syndrome"; Louise S. Kiessling, Anne C. Marcotte, Maggie Benson, Charles Kuhn, and D. Wrenn, "Relationship Between GABHS and Childhood Movement Disorders [abstract]," *Pediatric Research*, 33 (1993); Susan E. Swedo, Henrietta L. Leonard, Barbara B. Mittleman, Albert J. Allen, "Identification of Children with Pediatric Autoimmune Neuropsychiatric Disorders Associated with Streptococcal Infections by a Marker Associated with Rheumatic Fever," *American Journal of Psychiatry*, 154 (1997): 110–112.

16. A. J. Allen, H. L. Leonard, and S. E. Swedo, "Case Study: A New Infection-Triggered, Autoimmune Subtype of Pediatric OCD and Tourette's Syndrome," *Journal of the American Academy of Child and Adolescent Psychiatry*, 34 (1995): 307–311; A. J. Allen, "Group A Streptococcal In-

fections and Childhood Neuropsychiatric Disorders: Relationships and Therapeutic Implications," *CNS Drugs*, 8 (1997), 267–275.

17. Jean M. G. Itard, "Mémoire sur Quelques Fonctions Involontaires des Appareils de la Locomotion, de la Préhension et de la Voix," *Archives Générales de Médecine*, 8 (1825): 403–405.

18. Straus, "Über die organische Natur der Tics und der Koprolalie," pp. 698–699.

19. Laurence Selling, "The Role of Infection in the Etiology of Tics," *Archives of Neurology and Psychiatry*, 22 (1929): 1163–1171.

20. See discussion in Chapter 5.

21. For an overview of the questionable results of psychosurgery on Tourette's patients see Chapter 8 and Uta Asam and W. Karrass, "Gilles de la Tourette-Syndrom und Psychochirurgie," *Acta Paedopsychiatrica*, 47 (1981): 39–48.

22. Susan Hughes, *What Makes Ryan Tick? A Family's Triumph over Tourette Syndrome and Attention Deficit Hyperactivity Disorder* (Duarte, Calif.: Hope Press, 1966), p. 5.

23. Adam Ward Seligman and John S. Hilkevich, eds., *Don't Think about Monkeys* (Duarte, Calif.: Hope Press, 1992), p. 199.

Acknowledgments

Many persons have aided me in this project. I am grateful to my administrative assistants, Margaret Dennis and Kathy Holcomb, and my research assistants, Colin Talley, Erika Dressler, and Rebecca Frey. Kelly Martin and the staff at San Diego State University's interlibrary loan department tracked down hundreds of obscure references with amazing speed. Special thanks goes to my colleagues Stephen Roeder and William Hazen at San Diego State University; Michael Bernstein, Henry Powell, and Larry Hansen at the University of California, San Diego, and Paula Frew, now at Emory University. My interest in Tourette syndrome was sparked initially by Mitchell Medeiros, whose courageous lifelong battle to control unwanted tics and vocalizations continues today. Mitch's father Bill Medeiros has been extremely forthcoming, providing insights not available in a clinical setting.

My most extensive experience with patients has been at Memorial Hospital of Rhode Island and I thank all of them for providing me great insight and I hope empathy. My greatest debt at Brown University is to Dr. Louise S. Kiessling, who unselfishly shared her considerable knowledge of TS and clinical medicine, and Linda Abbott, RN, who allowed me to observe her great skill with patients. At Brown I also learned enormously from Drs. Joseph Hallett, Edward Brown, Anne Marcotte, and Julie Wilson.

Most of this book was written in Vancouver, British Columbia, where I spent 1996–1997 as the Jennifer Simons Professor of Graduate Liberal Studies at Simon Fraser University. The Simons Chair came with a stipend that supported research, travel, supplies, and much needed time away from my administrative duties. I received many useful suggestions and much stimulation from my colleagues Professors John Hutchinson, Sharon Fuller, and Harvey

Mitchell at SFU and at the University of British Columbia. My stay in Vancouver provided an opportunity to participate in the Child Neuropsychiatry Clinic at Children's Hospital, directed by the tireless and extremely knowledgable psychiatrist Roger Freeman. Here too, with prior permission and supervision, I observed patients with Drs. Freeman and Diane Fast, from whom I also learned much. I owe thanks also to my UBC medical school colleagues Drs. Bruce Bjornson, Elliot Goldner, Ismhael Laher, John Lively, Anton Scamvougeras, Morely Sutter, and Jim Wright for their many suggestions.

In Paris, Daniel Widlöcher, chief of the Psychiatry Clinic at the Salpêtrière, graciously opened many doors for me. Véronique Leroux-Hugon, the conservateur of the Bibliothèque Charcot, was extremely helpful as were the staffs of the Assistance Publique, Hôpitaux de Paris, and the archivists at the Hôpital Bicêtre. Also, I appreciate the conversations with and suggestions from Professors Yves Agid, Michel Dugas, and Jacques Poirier. Marc Lalvée, founder of the French Association for Compulsive Behaviors and Tourette's (A.F.T.O.C.), was extraordinarily generous in sharing information and helping me obtain materials for my research. Drs. Cyrille Koupernik and Pierrick Hordé are thanked for sharing their views on Tourette's.

I owe an unending debt to Sue Levi-Pearl, Medical and Scientific Liaison of the Tourette Syndrome Association in Bayside, Long Island, who played a role in almost every contact I have made. Betti Teltscher, a founder of the TSA, was extraordinarily helpful, even at a time of incredibly difficult personal tragedy. Elaine Novick generously shared memories and also her husband's, the late Dr. Sheldon Novick's, daunting correspondence with Tourette's researchers. Judi Wertheim, Erika Feinholtz, Abbey Meyers, and Cy Goldis answered my many questions about the history of the TSA. John Teltscher and Eleanor Pearl provided me with important insights on Tourette's.

This book benefited greatly from my conversations with the late Dr. Arthur K. Shapiro. Dr. Elaine S. Shapiro's many insights, excellent memory, and good humor have contributed immensely to my understanding of Tourette's and its history. Thanks also to Drs. Ruth Bruun and Arnold Friedhoff for filling me in on many details on the early years of the TSA and on their subsequent thinking on Tourette syndrome. Adele Lerner, archivist of the New York Hospital—Cornell Medical Center at New York-Cornell greatly assisted me by locating valuable patient histories. I also received advice, assistance, and encouragement from Coleen Wang, Susan Hughes, Adam Seligman, Dale Parent; Professors Stanley Finger, Ellen More, and Dora Weiner; Drs. David Comings, Mort Doran, Alan Rabin, and Mary Robertson. Professors Alice Wexler and Edward Shorter read my book manuscript for Harvard University Press. I am extremely grateful for their many constructive suggestions. My editor at Harvard, Joyce Seltzer, read every page with care, insight, and tough love. Thanks also to my production editor, Kate Brick, for her careful reading and suggestions.

I am responsible for all French translations. I thank my SDSU colleague historian Joanne Ferraro and University of Milan neurologist Claudio Luzatti for help with my Italian translations. Two graduate students, Nadine Meissen and Kerstin Stuerzbecher, helped with German translations. Some of the material in this book, though in different form, appeared in: "Medical Fictions: The Case of the Cursing Marquioo and the (Re)Construction of Gilles de la Tourette's Syndrome," *Bulletin of the History of Medicine*, 69 (1995): 224–254; "The Controversy over the Classification of Gilles de la Tourette's Syndrome, 1800–1995," *Perspectives in Biology and Medicine*, 39 (1996): 409–435 (written with Louise S. Kiessling); and "Freud and the Diagnosis of Gilles de la Tourette's Illness," *History of Psychiatry*, 9 (1998); 1–250.

The two illustrations of the basal ganglia were drawn for me by Bardy Anderson and are copyright © 1998 by Howard I. Kushner. The photograph of Georges Gilles de la Tourette is reproduced from Arthur K. Shapiro, Elaine S. Shapiro, Ruth D. Bruun, and Richard D. Sweet, *Gilles de la Tourette Syndrome* (New York: Raven Press, 1978) with the permission of Elaine S. Shapiro. Permission has been granted by the National Library of Medicine to reproduce the photograph of Jean-Martin Charcot. The photograph of Sandor Ferenczi is reproduced with permission from Paul Roazen, *Freud and His Followers* (New York: Knopf, 1972). The photograph of Margaret Mahler, by Trude Fleischmann, is reproduced from *The Memoirs of Margaret S. Mahler*, compiled and edited by Paul S. Stepansky (New York: Free Press), 1988, p. 89 with the permission of Paul S. Stepansky. Elaine S. Shapiro has granted permission to reproduce the photograph of Arthur K. and Elaine S. Shapiro from *Parade Magazine*, 15 October 1978. The photograph of Jack Klugman, Judy Wertheim, and Betti Teltscher is reproduced with the permission of Betti Teltscher.

This book was supported by several grants including a National Endowment for the Humanities, Science, and Technology Grant, RH-21165; a San Diego State University Research, Scholarship, & Creative Activity Fellowship; the Jennifer Simons Endowment of the Graduate Liberal Studies Program of Simon Fraser University; and the John D. Adams endowment at SDSU.

Carol R. Kushner, to whom I dedicate this book, has been a collaborator in every aspect of its production. A professional editor, she has edited the manuscript chapter by chapter, several times, and to the extent it is a readable piece, it is through her efforts. Carol has been a constant companion in discussing each idea or revision I have imagined. She has kept me from big and little errors and the book is as much ours as mine.

Index